EFFECTIVE
Business and Technical
PRESENTATIONS

THIRD EDITION

EFFECTIVE
Business and Technical
PRESENTATIONS

THIRD EDITION

GEORGE L. MORRISEY
THOMAS L. SECHREST

ADDISON-WESLEY PUBLISHING COMPANY, INC.
READING, MASSACHUSETTS ▲ MENLO PARK, CALIFORNIA ▲ NEW YORK
DON MILLS, ONTARIO ▲ WOKINGHAM, ENGLAND ▲ AMSTERDAM
BONN ▲ SYDNEY ▲ SINGAPORE ▲ TOKYO ▲ MADRID ▲ SAN JUAN

Library of Congress Cataloging-in-Publication Data

Morrisey, George L.
  Effective business and technical presentations.

  Bibliography: p.
  1. Lectures and lecturing.   I. Sechrest, Thomas L.
II. Title.
PN4193.L4M6 1987   651.7′3   87–1814
ISBN 0-201-15852-3

Cover design by Copenhaver Cumpston
Text design by Carson Design
Set in 11-point Meridian by Techna Type, Inc., York, PA

  BCDEFGHIJ-AL-898
Second Printing, March 1988

Contents

4 Handling Presentation Logistics   101

5 Delivering the Presentation   113

6 Conclusion   131

APPENDIX (REMOVABLE WORKSHEETS AND GUIDELINES)   137

Preface to Third Edition

Why a Third Edition?

There is an old adage that says, "If it ain't broke, don't fix it!" The second edition of this book continues to be one of Addison-Wesley's best-sellers even though it has been out for more than ten years. This is gratifying, of course, and reinforces our belief that the approach to preparing and delivering an effective presentation we advocate is still as sound as ever. This edition does not change that; it enhances it through more up-to-date examples. There have been some significant changes, however, in both the types and uses of presentation aids that will be addressed here. In addition, we discovered a need for more information on handling presentation logistics than was included in the first two editions. This has been expanded into a full chapter.

Although presentations continue to be a strong area of personal interest, most of my professional activity in recent years has been devoted to convention speaking, executive team consulting, and corporate seminars on management and organizational planning. While I still have a few clients with whom I work in the area of effective presentations, it represents a relatively small part of my current efforts. Consequently, I invited Tom Sechrest to join me as coauthor in this third edition. In addition to bringing a fresh perspective, Tom is especially knowledgeable about presentation aids and logistics and has had extensive experience in working with managers and key specialists from governmental organizations in improving their presentations. This book is much stronger as a result of his contributions.

What's Different about This Edition?

There is no conceptual difference in the basic approach that has stood the test of time. However, there are several other important changes that we feel will make this edition an even more valuable resource:

1. *Format.* The move to a larger format reflects the fact that the book's primary use continues to be as a working resource in formal training programs on effective presentations. The larger size makes it easier to complete the related practice exercises. Also, as in the second edition, numerous worksheets appear in the text, with an additional copy of each worksheet provided in the appendix, from which they can be removed and reproduced for ongoing use.
2. *New graphics.* In keeping with the contemporary style of this new edition, the graphics have been completely redone, thereby significantly enhancing the book's flow.
3. *Presentation aids.* Clearly the most substantial change has been in chapter 3, "Developing and Using Effective Presentation Aids." The state of the art in that technology has advanced rapidly since the last edition. This chapter provides the means for developing your presentation aids conceptually and describes a wide variety of practical techniques for translating the concepts into strong supporting tools for enhancing your presentation.
4. *Presentation logistics.* What was a section in the first two editions has been expanded to a full chapter in this one to reflect the growing recognition that the environment surrounding your presentation can be just as important as the presentation itself. Whether the logistical details related to the meeting are handled by someone else or not, it is in your best interest to make certain that nothing is overlooked that could have an impact on the outcome of your presentation.

We have also included the Preface to First Edition, which provides a comprehensive overview of the development of the book and how it can be used. It is just as valid today as it was then.

This is followed by the Preface to Second Edition. The alert reader may notice the omission from this third edition of certain material mentioned there. Two second edition appendix articles—on storyboarding, by Gus Matzorkis, and on video, by Tom Sechrest—have now been dropped in favor of covering these topics in the body of the text, given their increased relevance. And while the book retains its strong emphasis on Management by Objectives and Results (MOR) as a key to preparing and delivering effective presentations, the second edition's coverage of MOR as a separate topic has been dropped. In the years since the publication of the second edition, the principles of MOR have become so widely accepted that the average reader requires no introduction to them.

As always, we welcome your feedback on the value of the process and how it can be made even more effective.

Buena Park, California   G. L. M.
April 1987

Preface to First Edition

This Book Is Useless

unless you approach it with the idea, that, with careful thought and adequate preparation, an effective oral presentation or briefing can be made on any subject. This is not a collection of shortcuts and gimmicks that will make you a polished presenter. Nor is it a substitute for participation in a formal training program, although it can serve as an excellent text for such a program. It *will* provide:

1. A tested, step-by-step method that will result in a concise, interesting, and effective presentation.
2. Practice exercises, in connection with each of the steps, that you can do with the assistance of others, to ensure effective application of the suggested techniques.
3. Guidelines for a wise selection and use of audiovisual aids.
4. Instruction on the pitfalls you must avoid.
5. Insights for improving the communication relationship with your audience.
6. Techniques for increasing the effective use of your body and your voice.
7. Suggestions for making the best use of audience question periods.

Why Was This Book Written?

Actually, this book is the translation of a training program in briefing techniques that I first developed in 1962 at North American Aviation's Space Division in Downey, California. At that time, line management recognized the growing requirement for oral presentation of ideas in what has come to be known as the industrial briefing.

Such presentations were required from a wide variety of key management and technical personnel, most of whom had had little or no training or experience in making such presentations. It became increasingly apparent that a significant number of work-hours were being expended in preparing, in presenting, and even more critical, in listening to many briefings that were neither brief nor effective in conveying their message. Therefore, we were charged with the responsibility for developing a training program to remedy this situation.

The purpose of this program was not to develop polished public speakers. Many academic and commercial programs are already effectively doing this. Rather, its purpose was to provide the average manager or technical expert with a set of tools and guides for practicing so that he or she could give a *brief,* coherent, and reasonably successful oral presentation.

In the nearly five years since the course was developed, more than fourteen hundred people at North American have completed the program under several different instructors. Furthermore, there is a continuing waiting list of employees who want to participate in it on *their own time.* While there are no specific statistics on the subject, periodic postclass surveys of participants have indicated the following general results:

1. Reduction of from 20 percent to 50 percent in *briefing preparation time.*
2. Reduction of more than 50 percent in *presentation time* for the same type of briefing (before the program, average presentation time for many participants was an hour or more; after the program, average time for the same participants was twenty to thirty minutes).
3. Substantial reduction, without loss of effectiveness, in the number and complexity of charts used.
4. Significantly increased self-confidence about giving briefings.
5. Much better reception by, and corresponding effectiveness with, the people to whom the briefings were given.
6. Confirmation of these results by the superiors of many of these participants.

Several former participants in the program, as well as my fellow instructors, urged me to put the course into manual or textbook form. A survey of the field revealed that there were many publications on the subject of public speaking. However, there appeared to be none specifically designed as a "how to" for the relatively inexperienced person in industry, government, or business who must make a largely technical presentation to a critical audience. For these reasons, this book was written.

Needless to say, experience in conducting this training program at North American and elsewhere resulted in many modifications and an increasingly effective approach. I believe that this book captures the essence of the program in its current form. It can be most productive as a text in a formal training program with a skilled instructor. If no such instructor is available, the next best approach would be for a group to work together for mutual improvement. Having a skilled briefer in to lead the critiques and practice sessions would make the training more effective. However, if these suggestions are not practicable, an individual can use this text effectively as a self-teaching device. It is designed to be a practical, down-to-earth guide that will be used and reused regularly.

Who Can Benefit from This Book?

As mentioned earlier, this material is designed primarily for the relatively inexperienced presenter in industry, government, or business, not for the public speaker as such (although that person, too, could benefit from the ideas presented here). Typical of those who would find this a valuable guide are:

▲ *President* of a company, for a report to the board of directors or stockholders.

▲ *Sales engineer*, for a technical sales presentation to customer representatives.

▲ *Controller*, for an overview of the company's financial projections to a top-management group.

▲ *Manufacturing cost analyst*, for a review of staff loading requirements with the department manager.

▲ *Research scientist*, for presentation of the results of a study
   a. At a formal gathering of peers (for example, a national symposium).
   b. To management people not oriented to that technical field.

▲ *Industrial engineer*, for a review of a work-sampling study to the management of a less than fully cooperative department.

▲ *Credit manager*, for introduction of a new credit-application system to employees.

▲ *Governmental department head*, for presentation of an annual budget forecast to the appropriate legislative body.

▲ *Training instructor*, for presentation of a training lecture.

▲ *Employment specialist*, for an employee-recruitment presentation.

▲ *Project engineer,* for a report on the current status of a directed design change
   a. To his or her own management.
   b. To the customer.
▲ *Purchasing agent,* for a bid-seeking meeting with potential sub-contractors.
▲ *Supervisor,* for a motivational presentation to subordinates on workmanship.
▲ *Safety representative,* for an accident-prevention presentation to a group of maintenance supervisors.
▲ *Anyone,* with a requirement to present primarily technical information in a brief and understandable manner to a critical audience.

How Can This Book Be Used in a Training Program?

The personal preference and experience of the instructor and the particular circumstances of the moment will have considerable bearing on the approach to conducting the program. My own experience has proved successful when the following points were observed:

1. *Optimum group size*—12 to 15 persons

   I have conducted fruitful classes, however, with as few as 7 and as many as 25, in the latter case using a second instructor for divided practice sessions.

2. *Optimum program length*—21 to 27 hours

   It could easily be expanded to a full semester in a school situation, with greater subject depth and with more and longer practice presentations, with the following options:

   a. Weekly sessions of 2, 3, or 4 hours (3 usually seems best).
   b. A three-day intensive seminar-workshop.
   c. Combination of half- and full-day sessions.

   Note: program length is directly related to the number of participants because of the requirement for individual practice presentations.

3. *Preparation exercises*

   a. Preassignment—come to first session with a briefing topic (and general knowledge of subject matter) in mind.
   b. Write briefing objectives in class with small-group critique, instructor circulating.

    c. Prepare Preliminary Plan, in or out of class, with in-class small-group critique, instructor circulating.

    d. When practical and desirable, design charts, in or out of class, for instructor and group critique.

4. *Practice presentations by participants*

    a. At least two presentations of 10- to 15-minute duration (some variation up or down in duration is possible without loss of value).

    b. Tape recording (audio or video) for later instructor critique or self-critique by the presenter.

    c. Written and verbal critique by instructor and fellow participants.

5. *Instructor approach*

    a. Demonstrate variety of aids and approaches during formal presentations.

    b. Schedule practice presentations to start as quickly as possible after preparation material and related exercises have been covered, making initial assignments during first session.

    c. Discuss techniques related to "Delivering the Presentation" in short increments, interspersed as a change of pace between groups of practice presentations.

Again, let me emphasize that these considerations have proved effective for me in conducting these programs. Another instructor may be equally successful using a different combination. The material can be readily adapted to almost any approach.

How Can This Book Be Used by an Individual?

Recognizing that many individuals will want or, of necessity, have to use this book without benefit of an accompanying training program, I suggest the following approach for maximum benefit. As with most tools, the versatility and usefulness of this book will increase in direct proportion to the individual's effort and experience in using it. It will be of most value to you if you:

1. Skim through it quickly to get an overview.
2. Then, read it carefully, doing the recommended practice exercises.

3. Use it as a specific guide every time you make an oral presentation.
4. Refer to it for solving specific problems only after you are familiar with the total recommended concept.
5. Practice the recommended techniques every chance you get.
6. Start now!

Acknowledgments

There are many individuals to whom I am indebted for much of the material that is developed here. Over a period of several years, the hundreds of participants in my classes on this subject have really been the most help in refining and solidifying the approach. The feedback they have given me on the continued benefits they gained from this approach has provided me with the incentive to write this book.

Among many present and former coworkers who have made contributions to the content of this text, I want to express particular appreciation to:

▲ Albie Johnson, for the substantial help he gave me in developing the material on audiovisual aids in the original training program.
▲ Dave Lewis, my former supervisor, for his continued encouragement as well as for his critical content review.
▲ Gene Phillips, who served as an excellent devil's advocate on the manuscript draft and provided some specific ideas in the area of audience question periods.
▲ Tom Scobel, for his critical suggestions and encouragement during my first cut at writing it.
▲ John Traband, for his suggestions during the development of the original training program and particularly for his ideas on the use of the audience retention curve.

Finally, I want to offer my apologies as well as my thanks to Carol, Lynn, and Steve, my wife and children, for putting up with me during the hectic periods of writing and rewriting.

Downey, California   G. L. M.
August 1967

Preface to Second Edition

Why a Second Edition?

There are two fundamental reasons. First, and most obvious, is the need for an update. The book came out originally in 1968 and was a product of ideas and materials that were developed as early as 1962. There have been some changes in the state of the art (although, interestingly enough, not nearly as many as one might think). Also, there has been quite a bit of feedback from users of the first edition that indicates a need for a certain amount of "fine-tuning." This is further supported by my own experiences in conducting seminars and in observing and practicing the art of effective business and technical presentations.

Second, I want to tie it closer conceptually to Management by Objective and Results (MOR), the approach to management with which I have become identified. Interestingly, and to prove a point I make in my MOR seminars, I was practicing the approach long before I knew what it was. Perhaps that is one of the reasons why the original training program and the first edition of this book were so successful. The idea of *establishing objectives for the presentation* and then putting it together in such a way that they are accomplished is such a logical, commonsense approach that results-oriented people were naturally attracted to it. From there, the transition to MOR—which, from my admittedly biased point of view, is the only way to effectively manage anything—is equally natural.

What's Different about This Edition?

Conceptually, there is nothing different. The basic approach introduced in the first edition has stood the test of time. The differences are more in degree than in kind. Consequently, organizations that have been using the first edition will have no difficulty in making the transition to this one. Here are the principal differences:

1. *Elimination/reduction of male dominance.* Through the good-natured badgering of several of my feminist colleagues (notably Mary Fuller, Dru Scott, Doris Seward, and Theo Wells), I have consciously worked to eliminate male-dominant references. While I do not agree with some of the approaches advocated within the feminist movement, I must, in all good conscience, acknowledge the right of women to be looked upon and referred to as equals. If I have slipped on any such references in the new edition, it is a result of long-entrenched habit patterns and not by intent.

2. *MOR reference.* A new subheading, modified and additional copy in the text, and an appendix article have been inserted to show the natural relationship of this particular communications skill to the broader concept of Management by Objectives and Results (MOR).

3. *New worksheets.* These have been added for use in preparing a Preliminary Plan, Resource Material Selection, and planning the presentation. They are also included on perforated pages in the appendix, along with others, from which they may be removed and reproduced for continued use.

4. *Governmental references and illustrations.* In support of the first edition's wide use in the public sector, several references to and illustrations of governmental briefings have been added.

5. *Audiovisual aids expansion.* There are enough new developments in this area to justify a complete book in itself. I have attempted to highlight those tools and techniques that I feel will be of most use to the average presenter.

6. *Preliminary arrangements expansion.* This portion of the chapter on "Delivering the Presentation" has been expanded to include guidelines for various seating arrangements and a reproducible Preliminary Arrangements Checklist.

7. *Brief annotated bibliography.*

8. *Appendix articles.* In addition to the one on MOR identified above, there are two articles written especially for this book by experts in their fields. The first, "Storyboarding. . . For Briefings," by Gus Matzorkis, identifies a unique approach to getting meaningful involvement from others in generating and assembling ideas. The second, "Videorecording. . . For Briefings," by Tom Sechrest, deals with various applications of this relatively new communications medium.

Acknowledgments

My appreciation, as always, goes first of all to the many participants in my seminars who have forced me to continue working on improving the approach. In addition, I have received constructive feedback from many individuals and organizations that have used the first edition of the book, identifying "soft" and confusing areas that needed expansion or fine-tuning. I am particularly grateful to: Gus Matzorkis, not only for his fine article on storyboarding but for being a good friend and counselor at the right time; Reed Royalty, of Pacific Telephone, who attended one of my early public seminars on effective presentations and developed one of the most comprehensive in-company training programs on the subject I have seen, which has influenced many of the modifications included in the new edition; Jack Rush, of Rockwell International, my former training partner, for his continued constructive feedback on this and others of my management development efforts; and to Tom Sechrest, a young but rapidly rising star in the field of communications, for his excellent article on videorecording.

Buena Park, California   G. L. M.
October 1974

EFFECTIVE
Business and Technical
PRESENTATIONS

THIRD EDITION

1

Why Read This Book?

Presentations. You probably recall many you've sat through that went way over your head; that meandered with no beginning, middle, or end; that were an obvious ordeal for both presenter and audience. In short, you've undoubtedly witnessed your share of failures.

Now you've been asked to make a presentation yourself. Your immediate reaction? —*Panic!* What should I say? Where will I get the necessary information? How can I ever pull it all together? How can I make *my* presentation a success?

When trying to convey their ideas to others, so many managers and technical experts, extremely competent in their own fields, panic like this—completely forgetting the same kinds of analytical processes that make them experts in the first place. While, clearly, some presenters will always be more effective than others, you will increase your own presentation effectiveness, regardless of its current level, by following the relatively simple analytical approach described here. But while it may be "simple," that does not mean this approach will be "easy." It will require a discipline that may, in this context, be somewhat foreign to you. However, the potential payoff in satisfaction and accomplishment will be enormous.

In addition to satisfaction and accomplishment, enhancing the effectiveness of your presentations can have more tangible payoffs in terms of cost savings, for the formal technical presentation is one of the most expensive means of communicating information used by business, industry, and government today. There are, of course, the

1

obvious factors: the cost of planning and delivering the presentation (which can be substantial) as well as the cost represented by audience members' lost work time (far and away the most significant expense). Consequently, the organization sponsoring the presentation expects a worthwhile return on its investment.

But what of the presentation that fails to communicate anything significant? This failure adds to the already considerable expense for planning, delivery, and audience time a further, hidden cost that is virtually impossible to calculate.

Your first question, then, is whether or not a formal presentation is the most effective way to communicate your message. There may be acceptable alternatives that involve much less time, effort, and money.

When your analysis indicates that a formal presentation is indeed desirable, the process outlined in this book has been shown to produce cost savings (and minimize stress) by helping you reduce your preparation time by 20 to 50 percent, shorten the average presentation length by half, and keep your self-confidence during preparation and delivery. This, in turn, will enhance your reception and increase your effectiveness with the audience.

WHAT'S A PRESENTATION?

Everyone has a general idea of what's meant by a *presentation*. But let's define the term more specifically. A presentation (also called a *briefing* or an *oral presentation*) involves *the preparation and delivery of critical subject matter in a logical and condensed form, leading to effective communication.*

The components that contribute significantly to the effectiveness of a presentation are shown in figure 1-1. The presentation process has two major phases: *preparation* and *delivery*. In reality, there is no line of demarcation between these two phases: the best presentations are a continuum from start to finish.

The interrelationship between the components of an effective presentation can be summarized as follows:

▲ **Effective communication** is the goal of any presentation. Simply defined, this means getting the message across in a manner that will accomplish your objectives.
▲ **Audience results desired** must be identified in order for you to reach this goal. The importance of this consideration cannot

2

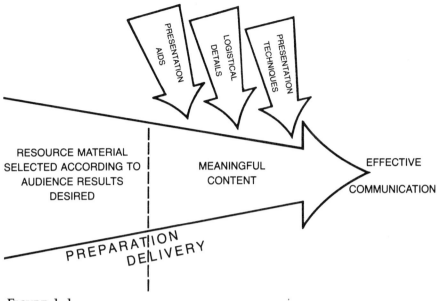

FIGURE 1-1
Elements of a Presentation

be overstated. The results desired will be the determining factor in your selection of appropriate resource materials and should be uppermost in your mind throughout both the preparation and delivery phases of your presentation.

▲ **Meaningful content** is at the heart of the presentation. Without this, by definition, you don't have a presentation. This meaningful content is supported by effective *presentation aids* and *presentation techniques* as well as by proper handling of *logistical details*. Understanding the *support role* of these elements is critical. If content is compromised in order to exploit unusual or available aids or gimmicks, employ a specific delivery approach (like a spontaneous live demonstration), or utilize a preferred but less than optimal meeting site, you might in the short run enhance audience acceptance of your presentation. In the long run, however, you risk not effectively communicating your message and thereby not accomplishing the presentation's real objectives.

And how do you know when you have achieved effective communication? If your audience does what you want it to do as a result of your presentation, you have accomplished your objective, even

3

though you might have violated every principle covered in this text in doing so!

TYPES OF PRESENTATIONS

Now that we've examined what a presentation is, we'll expand our definition by considering what types there are. Briefings are categorized according to their purpose, with those commonly used in business, industry, and government falling into four categories. While there is considerable overlap among these categories, and there may be elements of all four in a particular presentation, a presentation can, in general, be described as being one of the following:

1. **Persuasive.** Of course, every presentation is to a certain extent persuasive. First and foremost, you must convince the audience that you know what you're talking about. Beyond this, a presentation specifically designed to be persuasive might be used to:

   ▲ Pique the interest of a potential customer (or group of customers) in a new product, service, capability, or program that you are offering or will offer.

   ▲ Elicit an audience's confidence in the organization you represent and the message you are presenting.

   ▲ Convince upper management of the need to commit additional personnel or money or to utilize a particular methodology.

   ▲ Sell an existing customer on a new product, the modification of a current product or service, or a change in scope, funding, scheduling, or procedures.

   ▲ Persuade colleagues, employees, or members of parallel organizations to accept changes in operations or to acknowledge the need for closer coordination.

2. **Explanatory.** An explanatory presentation provides a general familiarization, gives the big picture, or describes new developments. The presenter's primary objective is to make new information available or refresh the audience's understanding of a given topic. Although an element of persuasiveness is present, this type of presentation requires a broad brush approach and rarely involves any significant amount of detail. It might be used to:

   ▲ Orient new employees to the organization.

▲ Acquaint staff members with what is involved in opening a new branch or division.

▲ Provide general information relevant to the needs of another department, company, or agency.

▲ Present information to professional associations, civic organizations, or other groups in the interests of good public relations.

3. **Instructional.** This type of presentation teaches others how to use something, such as a new procedure or piece of equipment. This objective usually requires greater involvement of the audience to reinforce learning and frequently requires detailed information. Typical uses for this kind of presentation are to:

▲ Instruct customers in the use of your services or equipment.

▲ Train customer representatives to instruct their own employees in the use of your services or equipment.

▲ Teach others to effectively follow specific procedures.

▲ Coach employees on specific behavior that will increase productivity.

4. **Oral report.** An oral report usually brings the audience up to date on something with which they are already familiar. Details may be provided on a selective basis, according to the needs and interests of the specific audience. Oral reports may be designed to:

▲ Update customer representatives on the status of a specific project.

▲ Inform management on current expenditures compared to budget.

▲ Provide a detailed accounting of the proceedings of a work group chartered to conduct an investigation or a research project.

HOW THIS BOOK IS ARRANGED

The arrows in the diagram (figure 1-1) shown earlier in the chapter represent the major inputs that directly affect the quality of a presentation: meaningful content, supported by effective use of presentation aids and presentation techniques and by careful attention to logistical details. The diagram also reflects the organization of this book.

Developing meaningful content is dealt with in chapter 2. It describes a six-step presentation-design model, each step of which is critical for planning the presentation's content. Overlooking any one of these steps may lead to serious flaws or gaps in the final presentation. Given that content is the foundation on which the effectiveness of a presentation is based, this chapter is the longest in the book and deserves the greatest attention.

The benefits and limitations of common presentation aids, and how they can be utilized to improve the effectiveness of a presentation, are dealt with in chapter 3. The visual sophistication of audiences and the increasing availability of new and often complex presentation aids requires that this subject be given special attention. Even though presentation aids clearly play a secondary role (that of supporting a presentation's content), the use of poorly designed aids or the failure to employ aids skillfully can destroy an otherwise good presentation.

Both chapter 2 and chapter 3 contain practice exercises to enhance the reader's understanding of the material. These exercises enable each reader to select a presentation topic that is relevant to his or her own needs and work actively on designing the presentation's content and developing its supporting aids.

The too-often-overlooked logistical details associated with a presentation are the topic of chapter 4, which outlines arrangements for everything from delivering a simple boardroom report to staging a major off-site event.

Presentation techniques, prime factors to consider in delivering your message, are covered in chapter 5. It describes platform skills, vocal techniques and problems, and ways to improve your interaction with the audience.

The bibliography will refer you to other resources that may be useful at various stages of the preparation and delivery process. Finally, the appendix contains reproducible copies of all worksheets and other aids found throughout the text. These valuable tools will quickly become an integral part of planning your presentations.

SUMMARY

The effectiveness of your presentations will be determined to a large extent by how closely you have followed the steps for proper preparation and delivery. Meaningful content, selected based on analysis of the desired results and supported by good presentation aids and platform skills, as well as by attention to the logistics of staging a presentation, will result in effective communication. This is not to imply that these steps are a panacea, but they do provide a logical and proven approach that should result in more concise and productive presentations. In short, if you manage your presentations with primary emphasis on objectives and results, the payoff on your investment can be tremendous.

2

Preparing a Presentation

Managing a presentation is no different from managing any other kind of investment. In allocating certain resources—primarily ideas, time, and personal energy, in this case—you expect to get a return that will exceed the value of the investment. In other words, you anticipate that your presentation, if successful, will bring about a result that has value for you. Defining the anticipated result and designing a presentation to accomplish that result is what this chapter is all about.

Careful thought during the planning stages may cause you to realize that the overall goal of effective communication might be accomplished more efficiently and productively by some means *other* than a presentation. Often a letter, telephone call, or technical bulletin will achieve acceptable results.

In considering a presentation as your communication approach rather than one of the aforementioned alternatives, you introduce a critical cost factor, involving both the cost of the presentation itself and the value of the *time* of everyone involved—presenter, support staff, and audience included! This significant cost factor must weigh heavily in your choice of a means to achieve your communication goal.

When your analysis indicates that a formal presentation is warranted, we suggest a six-step planning process (shown in the accompanying box) that involves breaking down its preparation into manageable units. These six steps are designed to help you produce a shorter, clearer presentation in less time and eliminate unnecessary

Steps in Preparing a Presentation

1. ESTABLISH OBJECTIVES for the presentation.

2. ANALYZE YOUR AUDIENCE.

3. PREPARE A PRELIMINARY PLAN for the presentation.

4. SELECT RESOURCE MATERIAL for the presentation.

5. ORGANIZE MATERIAL for effective delivery.

6. PRACTICE the presentation in advance and EVALUATE for necessary modification.

presentations. Each of these steps plays a vital role in the process, and none should be overlooked.

There are many presenters who already follow a number of these steps intuitively, employing something of a piecemeal approach to the preparation of presentations. This discussion is designed to concretely define the steps, and the interrelationship among them, and illustrate their practical application. This way, they can be followed as an *integrated system,* thereby providing presenters with the maximum benefit.

Following the discussion of your practical application project in the next section, we'll consider these six steps one at a time in the balance of the chapter, devoting a separate section to each.

PRACTICAL APPLICATION PROJECT

Since active involvement is a necessary condition for learning, practice exercises are suggested in this chapter for each of the six steps, with an additional group of practice exercises on presentation aids appearing at the end of chapter 3. At this point, you should select a topic that you might be required to make a presentation on in the near future and then, using that topic throughout, do the practice exercises. If it is easier for you to think in terms of an "artificial," or simulated, situation, here are some possibilities:

1. A sales presentation to a prospective customer.
2. A report to your supervisor and your peers on the current status of a major assignment for which you have responsibility.
3. A recruiting presentation to a group of prospects outlining the merits of joining your organization.
4. A presentation to a civic group on the mission and function of your organization.
5. A request for upper-management support in securing a budget increase for a particular project.

The more practical you can make your selected topic, the more valuable the practice exercises will be. Once you have chosen your topic, follow the six steps in order.

> **Steps in Preparing a Presentation**
>
> 1. **ESTABLISH OBJECTIVES for the presentation.**
>
> 2. ANALYZE YOUR AUDIENCE.
>
> 3. PREPARE A PRELIMINARY PLAN for the presentation.
>
> 4. SELECT RESOURCE MATERIAL for the presentation.
>
> 5. ORGANIZE MATERIAL for effective delivery.
>
> 6. PRACTICE the presentation in advance and EVALUATE for necessary modification.

———————— Step 1 ————————

ESTABLISH OBJECTIVES for the Presentation

Without a doubt, establishing objectives is the most critical step in the planning process. Surprisingly, it is the one most often overlooked. This step is designed to answer the question, "Why am I making this presentation?" *not* "What am I going to say?" You need to start by determining what it is you want to accomplish with a presentation. In other words, "What reaction do *I* want from the audience?" or "What do *I* want the audience to do as a result of this presentation?" These are short-term, self-centered objectives, which you may or may not share with your audience.

There is a natural tendency to confuse the objectives of what you are proposing in the presentation with the objectives of the presentation itself. THEY ARE NOT THE SAME! For example, in a new-product proposal to top management, an objective for the *project* might be "to obtain additional profitable business for the company." However, the objective for the *presentation* would be "to get approval to proceed with the project." An objective for a new job-training *program* in the Department of Labor might be "to provide unemployed citizens with the skills necessary to get and keep a job." An objective of a *presentation* on the program might be "to initiate action on research and development" or "to get a budgetary allocation for phase 1."

It is critical that you clearly understand the difference between the two kinds of objectives. The first stated objective in each of the two examples is a laudable long-range achievement and would undoubtedly be worth mentioning in a presentation. However, the presentation cannot, by itself, accomplish the long-range goal. It can only initiate action toward that goal. Consequently, you must identify what your presentation can accomplish, if successful, and focus your efforts on that end. If the short-term objectives for the presentation are not accomplished, the long-range objectives for the project or program are rather academic.

Presentation objectives give firm direction to the process of preparation. Instead of following the normal initial impulse to wade through reports, studies, charts, and so forth, much of which will be tossed aside later, you should invest your time at the outset in pinpointing the target toward which your efforts should be aimed.

Realistic Definition of Results Expected

It is important to keep in mind the *specific* results you expect from a presentation. These results should be:

1. **Realistic in scope,** so they can be accomplished in both the preparation time and the presentation time available. It is far better to do an effective job of presenting one major step of a program, opening the door for later presentations, than to try to cover the waterfront, so to speak, and run the risk of overwhelming the audience or doing a poor job on some of the steps. Making a presentation in well-prepared, comprehensible segments is often the key to acceptance of ideas.
2. **Realistic in view of the audience's knowledge and background.** Does the audience have the knowledge and background to achieve the results you want? Presenting the applications of a new computer in programming terms could be inadvisable for a group of production executives. A preliminary presentation on terminology might be in order or perhaps a broad, results-oriented presentation, with provision for later coverage of programming after audience members feel knowledgeable about the subject.
3. **Realistic in view of the audience's ability to act.** Do

audience members have the authority to make the decisions you would like? For example, if you are making a presentation on the "need for increased training in the company" to a group of first-line supervisors, it would be unrealistic to propose increasing the present training staff or adding facilities or equipment. They would not be authorized to make such a decision. However, proposing that they acquire greater familiarity with the capabilities of the training department, that they take time to train themselves and provide training time for employees, or that they give both verbal and written support to training efforts would be realistic.

4. **Realistic in terms of what you can reasonably expect to accomplish.** A costly plan of action may prove to be the ideal solution in a given situation. If, however, such factors as budget, personnel, floor space, or insurmountable resistance by key individuals will make acceptance of your proposal unlikely, it might be better to set as your objective now a "stripped-down" version of the plan that could lay the groundwork for gaining approval of the more costly solution at a later date.

Criteria for Judging Presentation Objectives

Presentation objectives should be specific and meet *one* or *more* of the following criteria:

1. They should answer the question, "Why am I making this presentation?"

   Example: *This is a regular monthly update to upper management on the status of the XYZ project. I want to enable audience members to spot any potential problems so that corrective action can be initiated.*

2. They should state the *results desired* from the presentation, in effect completing the sentence, "I want the following things to happen as a result of this presentation . . ."

   Example: *The customer will recognize the benefits of the new product and agree to make a purchase.*

3. If it is important to identify the body of knowledge to be presented, the objective should be qualified in terms of the results expected. "I want to tell about . . . so that . . . will take place."

Example: *I want to explain the new work process so that the employees on my team will both recognize its value to them in terms of increased efficiency and realize that it will not represent a threat to their jobs.*

Sample Objectives

Keeping in mind the requirements for presentation objectives, review the sample objectives in this subsection as models for your own.

Taking the "need for increased training in the company" as a principal topic, the objectives of the presentation might be:

1. To create an awareness of the need for increased training.
2. To gain management approval and support for increased training so that:
   a. Necessary funds will be authorized.
   b. Sufficient time will be authorized.
   c. Verbal and written support for training efforts will be forthcoming.

If you plan to give a presentation on the "support role of the financial department," the following might be an appropriate objective:

1. To explain the support functions of the financial department to line managers so they will:
   a. Recognize and use those services available to them.
   b. Accept guidance from financial department staff as desirable and helpful.
   c. Seek assistance before rather than after problems occur.
   d. Provide positive feedback that will enable the financial department to improve its services.

Secondary Objectives

In addition to a presentation's primary objectives, certain secondary results are often expected from a presentation that may not be shared with the audience. Identifying these results serves as a reminder to you that ancillary expectations can be met, even though they are

15

not communicated. Failure to openly disclose these secondary objectives is not meant to be devious. Rather, the achievement of objectives of this sort might be jeopardized by revealing them.

Some examples of secondary objectives are:

▲ *To establish personal credibility in this presentation, regardless of its outcome, so that future presentations will be favorably received.*

▲ *To share information with employees that may relieve anxieties over possible production cutbacks.*

▲ *To create awareness of the need to provide increased support for project ABC.*

▲ *To make the audience aware of and responsive to the other products, services, and capabilities offered by my organization.*

When specific, realistic primary and secondary objectives are established for a presentation, your "preparation energy" will be focused on achieving *results*.

Practice Exercises

1. Using the topic you selected earlier in the chapter, write a short, clear statement of objectives in line with the criteria outlined in step 1.

2. When you are reasonably satisfied with your objectives statement, ask some of your coworkers to critique it on its own merits—that is, with no comment or explanation from you. If, after reading the objectives once, they understand exactly what you are trying to accomplish with the presentation, you have successfully completed what is probably the most important, and perhaps the most difficult, step in the preparation process. If they raise objections that seem valid to you, rewrite your objectives in line with their suggestions, using your own good judgment.

Steps in Preparing a Presentation

1. ESTABLISH OBJECTIVES for the presentation.

2. **ANALYZE YOUR AUDIENCE.**

3. PREPARE A PRELIMINARY PLAN for the presentation.

4. SELECT RESOURCE MATERIAL for the presentation.

5. ORGANIZE MATERIAL for effective delivery.

6. PRACTICE the presentation in advance and EVALUATE for necessary modification.

Step 2

ANALYZE YOUR AUDIENCE

The first step, establishing objectives, focused on satisfying your *own* needs: determining what you, the presenter, want to accomplish. This second step, audience analysis, reverses the point of view. What do you need to know about the audience's knowledge, attitudes, likes, and dislikes in order to increase the probability of achieving your objectives? What is likely to get your audience to do what you want them to do?

Many an otherwise well-prepared presentation has fallen short of its objectives because the presenter failed to anticipate how the audience would react. For instance, one organization prepared what were basically good presentations for a group of top-level government officials. For the first four months, though, the reception from this group of officials was rather cool. The presenters were perplexed. The material was what the officials had requested. And the presenters had prepared what they thought were excellent visuals to illustrate the major points, using cartoon figures to build interest, a technique that can be very effective. However, in this case, the key official had a dislike for the use of cartoons. This personal idiosyncracy and the chilling effect it

17

had on the rest of the audience kept the presentation from achieving its objectives. A short preliminary outline of the presentation or a brief discussion of the format with one or two of the officials during the planning stage might have prevented this trouble. In any case, after the first presentation proved less than successful, a real attempt should have been made to determine why.

Although this second step is primarily used for advance planning, you must realize that it may be necessary to do some audience analysis on the spot and make modifications to your presentation based on data that were not available to you in advance. The more thorough your advance planning, though, the fewer on-the-spot adjustments you should have to make.

There are also many audiences for whom the type of detailed analysis described here would not be necessary. Regardless of the amount of detail required, you must give careful consideration to the audience's viewpoint as you plan a presentation.

Audience Analysis Audit (AAA)

One effective tool that can provide you with a picture of your audience and the best way to target your presentation is the Audience Analysis Audit, presented at the end of this section. This worksheet, when completed for a specific presentation you are planning, will help you identify audience characteristics and keep them in mind throughout the planning process. The AAA is divided into four major parts, each of which provides information for a specific use:

1. **Identify objectives for this audience,** which you did in step 1. Knowing what you want to accomplish as a result of the presentation will surface many of the specific audience characteristics you need to be aware of. Always keep the objectives in mind as you move through the planning process. Remember, a presentation on the same topic for two different audiences is likely to have different objectives for each.
2. **Analyze this audience in general terms,** to provide insight into the kind of *overall approach* most likely to achieve your objectives with this particular audience.
3. **Analyze this audience in specific terms,** to clarify the *scope* of the material. The analysis should indicate how deep into the subject it is advisable to go *from the point of view of the audience.*
4. **Select appropriate information and techniques.** The type of format and the specific approaches that would most likely

have a positive effect on this particular audience should be carefully chosen. (Better analysis in the area of techniques might have prevented the cartoon fiasco.)

At this stage in the preparation process, you should be alert for any additional relevant characteristics of your particular audience, not identified in the AAA, that you may also wish to investigate.

The information required to complete the AAA can be gathered in a variety of ways:

1. You can confer with persons who have made presentations to the same audience.
2. You can either confer directly with selected audience members or obtain your impressions indirectly by consulting with colleagues of audience members.
3. You can review examples of the work of audience members.
4. You can think the situation out logically and apply common sense to what you already know about the situation and the audience.
5. You can debrief after each presentation and assess the audience's reaction.

The important point is to take a systematic approach to analyzing your audience so that you not only avoid as many problems as possible but also plan the presentation most likely to accomplish your objectives with *this* audience. If you get into trouble with your briefing, it might be because you failed to pay enough attention to your AAA.

Practice Exercises

1. Using the Audience Analysis Audit as a guide, and supplying whatever additional information about the audience you feel would be helpful, write a brief profile or general description of the audience for whom you are preparing the presentation, using the topic you selected earlier.
2. Review your analysis or observations with a few of your co-workers to determine whether they would agree that your profile is as accurate as possible.
3. Identify in writing the specific audience for whom you are designing this presentation and give a short, one- or two-sentence summary of *pertinent* information about their knowledge, attitudes, and so forth.

Audience Analysis Audit (AAA)

(*Fill in the blanks or circle the most descriptive terms.*)

1. **Identify the objectives** of the presentation for *this* audience. What do you want the members of the audience to do as a result of the presentation?

   _____

   _____

   _____

2. **General analysis** of the members of this audience.
   a. What is their occupational relationship to you or to the organization you represent?

Customers	Top management	Public
Coworkers	Employees	Suppliers

   Other (describe): _____

   b. How long have they been in this relationship?

   _____

   c. What is their vocabulary understanding level?

Technical	Nontechnical	Generally high
Generally low	Unknown	

   d. How willing are the members of this audience to accept the ideas you will present?

Eager	Receptive	Neutral
Slightly resistant	Strongly resistant	Unknown

3. **Specific analysis** of the members of this audience.
   a. What is their knowledge of the subject?

High	Moderate	Limited
None	Unknown	

   b. What are their opinions about the subject and about you or the organization you represent?

Very favorable	Positive	Neutral
Slightly hostile	Openly hostile	Unknown

c. What are their reasons for attending this presentation?

_____

_____

_____

d. List some of the advantages and disadvantages of the presentation objectives to the members of the audience as individuals.

Advantages: _____

_____

_____

Disadvantages: _____

_____

_____

4. **Information and techniques**

a. What types of information and techniques are most likely to
gain the attention of this audience?

High-tech	Statistical comparisons	Cost-related
Anecdotes	Demonstrations	

Others (describe): _____

b. What information or techniques are most likely to get negative
reactions from the members of this audience?

_____

_____

_____

5. Briefly summarize the most important information from the preceding four sections.

_____

_____

_____

_____

_____

_____

_____

Steps in Preparing a Presentation

1. ESTABLISH OBJECTIVES for the presentation.

2. ANALYZE YOUR AUDIENCE.

3. **PREPARE A PRELIMINARY PLAN for the presentation.**

4. SELECT RESOURCE MATERIAL for the presentation.

5. ORGANIZE MATERIAL for effective delivery.

6. PRACTICE the presentation in advance and EVALUATE for necessary modification.

—————————————— Step 3 ——————————————

PREPARE A PRELIMINARY PLAN for the Presentation

There is another important step to take *before* you decide exactly how to put your presentation together. The Preliminary Plan is like a blueprint for the presentation. Its purpose is to build a framework on which the presentation can be developed and help you decide how much and what kind of material you will need. The Preliminary Plan is not designed to be a speaking outline but rather a conceptual guide to enable you to determine what will most logically lead to accomplishing your presentation objectives.

The Preliminary Plan serves two basic functions:

1. It forces you, the presenter, to carefully assess the direction to take, from selection of subject matter to keeping the flow of ideas channeled, and where to place emphasis for best results.
2. It establishes the parameters within which support personnel who provide backup data, prepare presentation aids, or assist in the presentation itself must work. Many of us have had the experience of assigning such responsibility to someone and

having the work done *all wrong* in terms of what we had in mind. The Preliminary Plan, if properly prepared, minimizes this problem because it gives both you and your support personnel specific *written* guidelines for completing the preparation process.

One caution: only a rare individual can make an effective presentation if that presentation has been entirely prepared by someone else. The person who will make the presentation must be actively involved in preparing the Preliminary Plan so that he or she will have provided direction for the concepts and approaches to be used. Only in this way will the final message reflect the presenter's own convictions, personality, and ability. Many an otherwise well-planned presentation has failed to produce results because it was given by someone who did not have a part in its preparation. If the presenter lacks sufficient time or motivation to contribute meaningfully to the preparation of the Preliminary Plan, perhaps it would be better for someone else to make the presentation.

How to Prepare a Preliminary Plan

The guidelines and accompanying worksheet that appear in this section are perhaps the *most* useful part of this entire text. Presenters who have based successful presentations on these guidelines often keep a copy of them in some prominent spot where they can readily refer to it.

Effectively using the guidelines on an ongoing basis will also enhance your ability to prepare a presentation on extremely short notice, even as little as twenty minutes—an important benefit!

The first and second guidelines, which reflect steps 1 and 2 of the planning process, serve as the foundation for the remaining guidelines. Guideline 3 is really the heart of the Preliminary Plan. It calls for stating the *main ideas* or *concepts* that the audience *must comprehend* if presentation objectives are to be accomplished.

Main ideas should be written as statements in conclusion form. The conclusions should be the ones you want the audience to reach about the material presented. These statements should *not* merely identify subjects to be covered, such as cost, scheduling, or organizational capabilities; *they should spell out what you want the audience to come to believe about these topics.* These may or may not be statements of fact.

Guidelines for Preparing a Preliminary Plan

The Preliminary Plan should be used as a guide:

▲ For the presenter in selecting materials, keeping ideas channeled, and determining emphasis points.

▲ For support personnel who may provide backup data, prepare presentation aids, or assist in the presentation itself.

1. **Identify specific objectives for the presentation,** keeping in mind one or more of the following criteria:
   a. They should answer the question, "Why am I giving this presentation?"
   b. They should state the results desired from the presentation—in effect, completing the sentence, "I want the following things to happen as a result of this presentation: . . ."
   c. If the body of knowledge to be presented must be identified in the objectives, use a sentence such as, "I want to tell about . . . so that . . . will take place."
   d. They should take into consideration any secondary objectives that you want to accomplish with the presentation.

2. **Identify the specific audience** for whom you are designing the presentation. State in a one- or two-sentence summary pertinent information about their knowledge, attitudes, and so forth.

3. **State the main ideas or concepts** that the audience *must* comprehend if the presentation objectives are to be met. These should:
   a. Be in conclusion form and preferably in complete sentences.
   b. Definitely lead to the accomplishment of the specific objectives.
   c. Be interesting in themselves or capable of being made so.
   d. Be few in number, usually no more than five.

4. **Identify necessary factual information** to support each of the main ideas and make them comprehensible to the audience. Avoid excessive detail.

However, they should provide guidance for the kinds of facts to be presented. Some examples:

- ▲ *The cost is reasonable and will be more than offset by future value received.*
- ▲ *The deadline can be met with effective management support.*
- ▲ *Engineering has the internal capability to perform the necessary design work.*

Frequently, there may be just one main idea in a presentation, even though it may be approached from several angles. You may wish to stress only the idea that "we will meet our primary objectives in spite of certain temporary setbacks," but do so in a number of different ways to make your point.

Only rarely should you try to present more than five main ideas if you want them to come through clearly and be remembered by your audience. If it is necessary to communicate a larger number of ideas in order to achieve some specific objective, it might be preferable to present them at two meetings, the first an overview or a request for input and the second a bid for a favorable decision. Consider the following example.

Sample Two-Session Approach

These might be the main ideas for explaining a proposed reorganization, which, in view of their number, are best presented on two separate occasions.

FIRST SESSION—MAIN IDEAS

1. Changes taking place in our primary market will have a significant impact on our profitability.
2. We are evaluating our organizational structure to determine ways of reducing overhead costs.
3. There will be no reduction in personnel as a result of any reorganization.
4. Based on the guidelines provided, each department head will make recommendations as to how best to reduce or contain costs.

SECOND SESSION—MAIN IDEAS

1. A decision has been made to contract with a management consulting firm for an analysis of cost reduction or containment alternatives.
2. Company A has been awarded the contract based on its experience with this type of assignment.
3. The efforts of the consulting firm will be coordinated through the XYZ unit.
4. Each department head will be responsible for evaluating the completed analysis.
5. Final recommendations will be made at the end of six months.

As you can see, the main ideas, as stated in the Preliminary Plan, may or may not be facts in themselves. They do not have to be. They must, however, be conclusions that can be reached on the basis of the factual information presented.

The fourth guideline requires identifying the *types* of factual information necessary to support and clarify the main ideas. Detail should be kept to a minimum unless the presentation is instructional or meant to provide specific information in some area with which the audience is already familiar. To continue the reorganization example, the following types of supporting factual information might be appropriate:

FIRST SESSION—SUPPORTING INFORMATION

Idea 1

Changes taking place in our primary market . . .

   a. Sales figures for several previous years, showing decreasing profits.
   b. Information about products offered by competitors, including retail costs and number of sales.
   c. Market projections by analysts, detailing expected trends.

Idea 2

We are evaluating our organizational structure . . .

   a. Overhead costs for several previous years, showing continual increases.

b. Estimated costs for future years if the present administrative setup is maintained.

c. Detailed cost figures for the units in which the largest overhead increases have taken place.

Idea 3

There will be no reduction in personnel . . .

a. Units where employees might be affected and how these employees could be reassigned.

b. Plans for retraining affected employees.

Idea 4

Based on the guidelines provided . . .

a. Areas in which costs might be reduced or contained.

b. Nature of recommendations expected from each department.

SECOND SESSION—SUPPORTING INFORMATION

Idea 1

A decision has been made to contract with a management consulting firm . . .

a. Reasons for the decision.

Idea 2

Company A has been awarded the contract . . .

a. Reasons for the selection.

b. Background information on company A.

Idea 3

The efforts of the consulting firm will be coordinated . . .

a. Rationale behind the choice of the XYZ unit as coordinator.

b. Procedures for the work to be conducted by the consulting firm.

Idea 4

Each department head will be responsible . . .

    a. Suggested evaluation procedures.
    b. Key indicators to look for.

Idea 5

Final recommendations will be made . . .

    a. How recommendations will be evaluated.
    b. Implementation timetable.

You can easily see from this example how necessary the Preliminary Plan is to the successful accomplishment of presentation objectives. To usefully compile the elements of the plan, you may wish to use the Preliminary Plan Worksheet, which follows the practice exercises. The worksheet will help shorten the time it takes to prepare a Preliminary Plan and ensure that you consider all the necessary elements. Three sample Preliminary Plans are shown after the worksheet.

Practice Exercises

1. Write a Preliminary Plan, following the Guidelines for Preparing a Preliminary Plan, for the presentation topic you have selected. You may wish to use the Preliminary Plan Worksheet. Concentrate in particular on guideline 3, the main ideas or concepts. Be certain that the main ideas are stated in the form of *conclusions* you want the audience to reach and are not merely a list of subject areas. These statements should, in most cases, be complete sentences.
2. Ask some of your coworkers to critique your Preliminary Plan, particularly the *main ideas*. Find out whether the statements are clear to them; ask for their suggestions on clarifying the meaning. See whether they feel that the main ideas meet the criteria established in the guidelines and that these ideas will accomplish your objectives. Determine whether your coworkers approve of your choice of supporting factual information.

Ask for recommendations. Do not feel, however, that you must justify your own approach. Reactions may be very helpful but, ultimately, you are the one who has studied and given careful thought to the content. Listen to comments and use your own judgment about how much weight to give them during the planning process.

Preliminary Plan Worksheet

Topic of the presentation: _____

_____

Approximate date, time, and place for the presentation: _____

_____

Who requested that the presentation be made? _____

_____

Presentation objectives (what will be the immediate results if the presentation is successful?):

1. _____

_____

_____

2. _____

_____

_____

3. _____

_____

_____

4. _____

_____

_____

Audience analysis (who are they, and what is their general knowledge of, interest in, and attitude toward the subject?):

_____

_____

_____

Main ideas or concepts that the audience must comprehend and retain if the presentation objectives are to be met:

1. _____

2. _____

3. _____

4. _____

5. _____

Factual information necessary to support the main ideas:

Idea 1

_____

_____

_____

_____

Idea 2

_____

_____

_____

_____

Idea 3

_____

_____

_____

_____

Idea 4

_____

_____

_____

_____

Idea 5

_____

_____

_____

_____

Preliminary Plan: Sample 1

Topic:

Progress report on XYZ project.

Objective:

To keep the customer informed on a regular basis about the status of the XYZ project; to communicate any problem areas and explain corrective measures being taken.

Audience:

Customer project director and related staff personnel. They are familiar with the project and will be interested primarily in the proposed delivery date.

Main ideas:

1. XYZ project is currently eight weeks behind but can be brought back on schedule with the following adjustments:
   a. Elimination of a redundant testing procedure.
   b. Minor change in product packaging.

2. Cost will remain within budget if these adjustments are made.

3. Project can be completed within three weeks if adjustments are made now.

4. Performance standards are all being met.

Factual supporting information:

Main idea 1

XYZ project is currently eight weeks behind . . .
   a. Factors causing the delay.
   b. Production changes being recommended.

Main idea 2

Cost will remain within budget . . .
   a. Detailed savings to be realized from production changes.

Main idea 3

Project can be completed within three weeks . . .
    a. Detailed time estimates until project completion.

Main idea 4

Performance standards . . .
    a. Key performance measurements.

Preliminary Plan: Sample 2

Topic:

Need for increased training in the company.

Objectives:

1. To create an awareness of the need for increased training.

2. To gain management approval and support for increased training so that:
    a. Necessary funds will be authorized.
    b. Sufficient time will be authorized.
    c. Verbal and written support for training efforts will be given.

Audience:

Top management plus other management personnel at Director level or higher. Most will have a general knowledge of the subject. A few will be favorably inclined to increase training, but most will be neutral, skeptical, or slightly hostile.

Main ideas:

1. Increased training is essential if we are to survive in the industry.

2. Money invested in training now (charged to overhead or taken from profit) will be returned manyfold in the future.

3. Time spent in training now (taken from current work) will result in a much more profitable use of time in the future.

Factual supporting information:

Main idea 1

Increased training is essential . . .
   a. New technology requirements.
   b. Training experience in similar companies.
   c. Potential application of new management concepts.

Main idea 2

Money invested in training . . .
   a. Recent training progress in the company.
   b. Comparative cost-of-operation figures (before and after training).
   c. Training cost versus the cost of replacing personnel.

Main idea 3

Time spent in training . . .
   a. Comparative (before-and-after) time-investment ratio.
   b. Intangible time benefits (for example, increased confidence and competence on the part of trained personnel, resulting in more productive use of time).

Preliminary Plan: Sample 3

Topic:

Support role of the financial department.

Objective:

1. To explain the support functions of the financial department to line managers so they will:
   a. Recognize and use those services available to them.
   b. Accept guidence from financial department staff as desirable and helpful.
   c. Seek assistance before rather than after problems occur.
   d. Provide feedback that can enable the financial department to improve its service.

Audience:

Division line managers. Most will have limited knowledge of or interest in financial department functions and will be neutral to slightly hostile.

Main ideas:

1. The financial department is in business to help make the line managers' job less complicated.

2. The financial department makes many useful services available to line managers in addition to the familiar, routine operations (payroll, financial reports, and so on).

3. Financial department personnel are skilled professionals with a desire to help.

4. Early identification of potential problems can enable more useful and less costly assistance to be provided to line managers.

5. The financial department is continually seeking feedback from line managers on ways to improve service to them.

Factual supporting information:

Main ideas 1 & 2

The financial department is in business . . .
The financial department makes many useful services available . . .
    a. Brief identification of *only* those services of direct concern to *this* audience.

Main idea 3

Financial department personnel are skilled professionals . . .
    a. Specific examples of services that have been or could be provided—in terms of benefits to the user (line managers).
    b. Brief identification of key staff members likely to relate to *this* audience.

Main idea 4

Early identification of potential problems . . .
    a. Specific examples of positive consequences of early identification. (Avoid mentioning, or minimize negative consequences of, failure to identify potential problems early.)

Main idea 5

The financial department is continually seeking feedback . . .
    a. Specific examples of improvements made as a result of feedback.

Steps in Preparing a Presentation

1. ESTABLISH OBJECTIVES for the presentation.

2. ANALYZE YOUR AUDIENCE.

3. PREPARE A PRELIMINARY PLAN for the presentation.

4. **SELECT RESOURCE MATERIAL for the presentation.**

5. ORGANIZE MATERIAL for effective delivery.

6. PRACTICE the presentation in advance and EVALUATE for necessary modification.

_____ Step 4 _____

SELECT RESOURCE MATERIAL for the Presentation

For most presentations, finding *enough* resource material to include is not difficult. The problem, rather, is one of proper selection, of determining what and how much material should be included.

If you have done your homework on the first three steps, step 4 becomes relatively easy. Rather than plowing through a mass of material, as might have been your natural inclination at the outset, you proceed in a much more focused manner. Now that you've reached this point, it is primarily a case of applying common sense, asking yourself a series of logical questions as you complete the task of determining what material should be included.

37

Questions for Guidance

The proper preparation of a Preliminary Plan is basic to the effective selection of resource material. The questions below will largely follow the Preliminary Plan.

1. What is the **object** or **purpose** of the presentation?
   Is it to be persuasive, explanatory, instructional, or an oral report? Do you want to arouse interest, test an idea, recommend action, inform, or resolve problems? (Review the objectives in your Preliminary Plan.)

2. What should be **covered?** What can be **eliminated?**
   Supporting factual information identified in the Preliminary Plan should indicate the subject matter to be covered. Unless items contribute significantly to the accomplishment of presentation objectives, they should be eliminated.

3. What amount of **detail** is necessary?
   The amount of detail depends on several factors: preparation and presentation time, the audience and its particular interests, and how much the audience must know in order to accomplish presentation objectives. Most presentations we have observed include much more detail than necessary. It is far better to leave the audience a bit hungry, wanting more detail, than to give them so much that they get confused or bored. You may wish to have additional details available, in the event that you are asked, without including them in the actual presentation. Supplementary details can always be made available in a handout, as well.

4. **What must be said** if the presentation objectives are to be reached?
   The answer to this question depends on the main ideas identified in the Preliminary Plan. You must decide what specific resource material is essential if the main ideas are to be accepted by the audience.

5. What is the **best way to say it?**
   Primarily considering the audience, what types of subject matter and what method of presentation (examples, anecdotes, statistics, comparisons, and so forth) do you feel will be most effective in getting the main ideas across?

6. What kind of **audience action** or **response** is required if the objectives are to be met?

   Do you need to force an immediate response (approval of a plan, authorization of additional money, and so on)? Or should you provide food for thought that will establish a favorable climate for later follow-up?

7. What material should be **withheld from the presentation** itself but be **available for reference** if required?

   Is there some information that is not essential to the presentation objectives but that you should have in reserve in the event questions about the topic are raised by someone in the audience?

8. Finally, submit all resource material to the **"Why?"** test.

   Try to look at the material objectively, as a disinterested observer. Examine each item selected for inclusion in the presentation and ask yourself, "Why is this to be used? What contribution will it make to achieving presentation objectives?" Whatever cannot withstand this critical evaluation should be eliminated. Answering this question can be somewhat painful, because there is a natural tendency to include information that is especially interesting or meaningful to you. But, in reality, this information might hold very little interest for the audience and, worse, might dilute the ideas essential to accomplishing the presentation objectives.

Are these questions a magic formula for the proper selection of resource material? No! Do they involve the application of common sense? Yes! The logical and careful analysis of material to be selected for inclusion in a presentation, always with the audience and the objectives in mind, is vital to effective preparation of the presentation.

Practice Exercises

1. Using the topic you have selected, answer the questions in step 4, identifying in writing all the resource material you will include in your presentation. Make a list of supplementary material you may want to have available for reference if requested.
2. Have coworkers review your selection. You should be able to justify the inclusion of each item, both to them and to yourself, or else eliminate it.

Steps in Preparing a Presentation

1. ESTABLISH OBJECTIVES for the presentation.

2. ANALYZE YOUR AUDIENCE.

3. PREPARE A PRELIMINARY PLAN for the presentation.

4. SELECT RESOURCE MATERIAL for the presentation.

5. **ORGANIZE MATERIAL for effective delivery.**

6. PRACTICE the presentation in advance and EVALUATE for necessary modification.

Step 5

ORGANIZE MATERIAL for Effective Delivery

Once you have selected resource materials in line with your Preliminary Plan, they must be organized into an effective presentation that will reflect your abilities, convey your honest beliefs, meet your objectives, and satisfy the needs of your audience. This step is where we get into the specific presentation outline and approach that are most likely to achieve the objectives you have established.

We will start by considering the characteristic changes in the level of audience attention—and hence, retention—during the course of a presentation, which will strongly affect your choices regarding timing and emphasis. And in terms of structure, we will examine the introduction, body, and conclusion—the three major parts into which most presentations can be broken down. Each serves a very specific purpose and requires a distinct approach.

Audience Retention

What your audience remembers determines whether or not you achieve your presentation objectives, so you need to be aware that

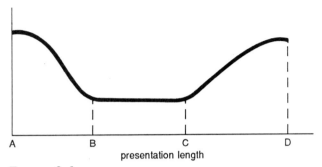

FIGURE 2-1
Audience Retention Curve

certain parts of a presentation are remembered better after the presentation is over. Figure 2-1 represents the *audience retention curve*. While many factors—such as time, room arrangement, predisposition of the audience, and so forth—have an effect on this curve, it does tell us something about how an audience will react to a presentation.

Curiosity will probably result in a reasonably high level of both attention and retention during the introduction (A). The retention level for a bad introduction, however, will be just as high as for a good one and will color the audience's attitude toward the entire presentation. This fact makes it doubly important to do a good job of gaining audience support during the introduction.

Following the introduction, a reasonably sharp drop in retention (B) is to be expected, even with a good presentation. Unless the content is very short and simple, a drop *will* occur, of that we can be sure. The bottom of the curve is not as smooth as the illustration would indicate. There are peaks and valleys during the body of the presentation. A deliberate change of pace, a story, an illustration, or audience activity during the body of a presentation can help counteract this drop in retention, although they can't completely overcome it. Repeating main ideas throughout the body of the presentation, and highlighting them succinctly in the conclusion, is about the only way to make sure the message gets across.

The rise at the end of the curve (C) depends on the effectiveness of the presenter. The presenter who runs out of things to say and simply stops may end the presentation with the retention curve at the low level indicated by point C. With a presenter who runs out of things to say and *doesn't* stop but instead rambles on with unnecessary repetition, the curve will drop off to nothing or, worse, may rise, indicating that the audience will go away remembering the dull ending and

41

thinking the entire presentation was a waste of time. If the presenter uses an appropriate summation-type statement to tip the audience off that the presentation is coming to a close, however, the retention curve will take an upward swing, and the effective presenter can take advantage of this rise in attention to help those in the audience who may have been mentally wandering to get back on track and carry something of value away from the presentation (D).

Introduction

The introduction has three primary purposes. The first and perhaps most important is to *sell the audience on listening to your presentation.* The introduction is a particularly critical time in a presentation. With most audiences, as noted earlier, there is a natural curiosity at the beginning. Whether or not the audience retains that curiosity or interest will depend to a large extent on your opening remarks and how effectively you convince them that it will be worth their while to listen attentively to what you have to say.

The second and more obvious purpose is to *introduce the subject* or reason for the presentation. A simple, accurate statement is called for. This part of your introduction should be interesting and *brief!*

The third purpose is to *establish your personal credibility* with the audience. Tell them why *you* are the one making the presentation. If another person is going to introduce you, give him or her information that will help establish your credibility.

SIX SAMPLE INTRODUCTIONS

The specific approach you use to introduce a presentation will depend on the subject matter, the time available, the audience, and your personal preference. Here are six of the many different ways in which a subject might be introduced. For illustration, the "need for increased training in the company" will again be used as our topic.

1. **Direct statement** of the subject and why it is important to the audience.

   Ladies and gentlemen, we are at a particularly critical time in the life of our company. In order to advance or even to survive in this highly competitive industry, we must substantially upgrade our performance capabilities. There are two ways in which we can do this: we can hire the necessary talent from the labor market, or we can train the talent we already have to meet the new challenges. Experience has shown that the first way is prohibitively expensive. Furthermore, recent surveys indicate that such talent is extremely scarce in our area. The second alternative—a systematic, well-planned program of ongoing training for our current personnel—appears to be the logical one for meeting this critical need. Let's examine the factors that lead us to this conclusion.

2. **Indirect opening** dealing with some *vital interest* of the audience and connecting presentation objectives with that vital interest.

   How much time did you spend last week correcting or redoing the work of some of your employees because they lacked sufficient knowledge or understanding of what was expected? One hour, two hours, five hours, more? Could you have used that time more productively by working at your own level, developing new business, or perhaps spending a leisurely evening at home with the family? How much of your time and effort might have been saved if your employees were adequately trained? How much more time and effort will you have to expend in the future if steps are not taken today to meet the rapidly growing demand for increased technological and managerial knowledge and skill? How will an expanded company training program give you greater flexibility and allow you to use your time more effectively and efficiently? Let's see!

3. **Vivid example** or comparison leading directly to the subject.

   Last month our company had to forgo bidding on a potential $10 million contract because analysis showed that, under present operating conditions, we do not have a sufficient number of trained personnel to handle the production, nor could we hire and train them fast enough to meet the time requirements of this contract. How many such contracts can we afford to lose to our competition? The time to act is now. This means instituting a far more comprehensive training program.

4. **Strong quotation** related to the subject that will be particularly meaningful to the audience and establish some rapport between you and them.

   When this company was established, our founder stated that its policy would be to utilize whatever resources were necessary to prepare employees to move with the organization as it changed, developed, and grew.

As we assess the impact of new technology on our work, we can see that training our employees to maintain our competitive edge is critical. Today, we will concentrate on the role this training will have in preparing our organization for the future.

5. **Important statistics** related to the subject.

The cost of obsolescence and depreciation of facilities and equipment is a necessary expense we know we must plan on. Last year, in our company, depreciation amounted to some 10 percent of the value of our capital assets.

In our business, trained personnel are also a major asset, and we must realize that a comparable allowance should be made if we are to maintain a labor force that is up-to-date, informed, and knowledgeable about all the advances being made in production. It is extremely important that we do this if our company is to grow and gain a larger share of the market. Yet our investment in terms of training hours per hours worked was only 0.4 percent last year. While other factors are obviously related to human depreciation costs, these facts do raise some interesting questions about whether we are making a sufficient investment in training our people effectively to meet tomorrow's challenges. Let's examine this question in greater depth.

6. A **story** or **anecdote** illustrating the subject, provided it is directly relevant to the presentation and is not merely contrived for entertainment purposes.

A classic space-age cartoon shows a spaceship on its way to some distant planet with the plaintive question emerging from somewhere inside, ''What dya mean, it won't flush?'' Now this appears to indicate insufficient training on the part of someone—the astronauts, the assemblers, or the engineers. So that we won't get caught with our *pants down, let's examine our* own *training requirements.*

Any one of these six introductory approaches might be appropriate under a certain set of circumstances and inappropriate under a different set of circumstances. Since the image you project at the outset is of vital importance in getting the audience's attention, you should open your presentation using whatever approach you can handle most effectively. Let's face it! You might not be a great storyteller. Therefore, for you, opening with a story or an anecdote might be a poor choice. Normally, you select just one approach for the introduction, bearing in mind that you want to sell the audience on listening to your pres-

entation as well as introduce the subject and establish your own credibility.

Body

The main ideas, once stated, need to be explained in whatever
detail is necessary to accomplish presentation objectives. Often, the
audience will look for patterns or relationships among the main points.
The ideas can be developed by using a problem/solution approach, a
cause/effect approach, a chronological approach, or some other
method that helps the audience understand the subject. The approach
should highlight their particular interest in certain aspects of the subject
and take into consideration the collective likes and dislikes with respect
to certain methods of presentation identified through the Audience
Analysis Audit. The proper use of illustrations is so important that, if
a choice must be made, it is usually better to eliminate some specific
facts than to cut out illustrations of those facts that are more critical
to the accomplishment of your objectives. You may wish to use some
of the following approaches:

▲ *Examples* illustrating or describing the ideas in operation. These
  include flowcharts, anecdotes, and definitions of ideas and approaches.
▲ *Reiteration* of the main ideas in the same or different words to
  help summarize the points, drive them home, and ensure that
  the audience remembers them.
▲ *Statistics,* if used sparingly, presented as simply as possible, and
  supported by visuals. It is not easy to argue with the facts if
  they are backed up by convincing statistics.
▲ *Comparisons* and *analogies* that place new ideas in a more familiar light.
▲ *Expert testimony* from users of the product or procedure and
  others who can address the subject area with a degree of credibility that is acceptable to the audience.

The main ideas should be logically sequenced so that they make
sense to the audience. If the subject matter is controversial, begin with
the least-disputed idea and carefully work your way toward those
concepts that are most resisted by the audience.

Conclusion

Now let's review the main points we've covered.
To sum up these factors . . .
Our primary purpose today has been to . . .
Reflecting on what we've discussed . . .

Using such a phrase makes it quite evident that you are going to wind things up. This has a tendency to bring the members of the audience back on target and gives you an opportunity to restate the main ideas and suggest agreement or recommend action.

Any idea you want the audience to remember needs to be repeated *from three to ten times* during a presentation, either verbatim or expressed in different words with a slightly different slant. Repetition, if handled intelligently, does not insult your audience, despite the common belief that it does. Obviously, you must exercise good judgment in your use of repetition, but you are naive indeed if you expect the majority of your audience to remember any point, even a significant one, if you make that point only once.

Since the conclusion tends to be the weakest segment of most presentations, it deserves as much planning as the rest of the presentation, if not more. The conclusion can be relatively brief, but it should be vivid and to the point in terms of what you want the audience to carry away with them, and it should be delivered in an upbeat manner. In effect, it should be a minipresentation that captures your primary message.

Proper attention during the planning process can ensure that the organization of the material in your presentation is purposeful. The most effective structure involves an *introduction* (which sells the audience on listening to your presentation, introduces the subject, and establishes your personal credibility), a *body* (which makes up the bulk of the presentation and provides the detail necessary for audience comprehension), and a *conclusion* (which allows you to summarize the main ideas, review the purpose of the presentation, and appeal for audience action).

Use all the imagination at your command and keep your specific objectives in mind as you devise effective methods for presenting your material.

The Guidelines for Organizing Material for Presentation and the Presentation Worksheet, both found at the end of this section, may be helpful to you in organizing your material. The worksheet provides a means for coordinating time, content, and presentation methods within the context of your objectives.

Practice Exercises

1. Using the topic you selected, organize the material for your presentation, following the Guidelines for Organizing Material for Presentation and using the Presentation Worksheet. Prepare the outline in rough form so that it can be changed if necessary.
2. Have your coworkers look over your worksheet, give you their reactions, and suggest ways to increase your presentation's effectiveness. Carefully consider these comments, but remember that it is *your* presentation and you should use those materials and approaches that will work best for you.

Guidelines for Organizing Material for Presentation

Introduction

The introduction has three primary purposes: (1) selling the audience on listening to the presentation, (2) introducing the subject matter, and (3) establishing your personal credibility.

Suggested approaches for the introduction:

1. *Direct statement* of the subject and why it is important to the audience.

2. *Indirect opening* dealing with some vital interest of the audience that can be linked to the subject.

3. *Vivid example* or comparison leading directly to the subject.

4. *Strong quotation* related to the subject.

5. *Important statistics* related to the subject.

6. *Story* or *anecdote* illustrating the subject.

Identify the methods you will use to state the idea and purpose of your presentation.

Body

Following the *main ideas* listed in the Preliminary Plan, provide the necessary detail for audience comprehension. Use examples, reiteration, statistics, comparisons and analogies, and expert testimony as methods to present material.

Visual illustrations are one of the most important aids to support the content. Plan carefully for their use. Also, indicate how questions and/or group discussion will be handled.

Conclusion

The conclusion is critical. It provides a summary of the main ideas, a review of the purpose of the presentation, and an appeal for audience action. It is, in effect, a minipresentation in and of itself.

Presentation Worksheet

Presentation topic: _____

Presenter(s): _____

Date, time, place: _____

General Considerations

1. How will the room be arranged (seating, name cards, and so on)?
   _____
   _____

2. How many are expected to attend? How and when will they be notified of the presentation? _____
   _____
   _____

3. What presentation aids will be required? Will equipment be available at the presentation site, or must someone transport it there?
   _____
   _____
   _____

4. Will handouts be used? What arrangements have to be made for them? How and when will they be distributed? _____
   _____
   _____
   _____

5. How and when will you handle audience questions? _____
   _____
   _____
   _____

Presentation Outline

Time allotted	Content*	Methods, aids, examples

Introduction: sell the audience on listening, introduce the subject, establish personal credibility; *body:* develop the main idea(s); *conclusion:* summarize content, appeal for action.

Presentation Outline

Time allotted	Content*	Methods, aids, examples

Introduction: sell the audience on listening, introduce the subject, establish personal credibility; *body:* develop the main idea(s); *conclusion:* summarize content, appeal for action.

Steps in Preparing a Presentation

1. ESTABLISH OBJECTIVES for the presentation.

2. ANALYZE YOUR AUDIENCE.

3. PREPARE A PRELIMINARY PLAN for the presentation.

4. SELECT RESOURCE MATERIAL for the presentation.

5. ORGANIZE MATERIAL for effective delivery.

6. **PRACTICE the presentation in advance and EVAL- UATE for necessary modification.**

Step 6

PRACTICE the Presentation in Advance and EVALUATE for Necessary Modification

It's a rare individual who can take even a well-prepared presentation and deliver it effectively on the first attempt. Most of us have had the experience of planning a presentation that looks good on paper only to have it fall flat in the real world.

Here are some presentation pitfalls (probably corollaries of Murphy's Law) that turn up regularly:

1. The spoken words don't flow as smoothly as they seem to on paper.
2. You lose track of where you are because of some distraction.
3. The mechanics of handling presentation aids interferes with your pacing.
4. You discover you are not as knowledgeable about the subject matter as you had originally presumed.

5. Someone in the audience asks a question you did not anticipate or are not prepared to answer.
6. The audience is cold and unresponsive.
7. The location selected does not lend itself to the type of presentation planned.

Rather significantly, all but the last two pitfalls can be directly attributed to insufficient preparation or practice. Even the last two can be dealt with if some foresight is used during planning, preparation, and practice.

Practice does not ensure success, nor will it make a good presentation out of a poorly prepared one. What practice *can* do is:

1. Increase your self-confidence and poise, making the audience more willing to believe you.
2. Reveal flaws or gaps in your material.
3. Give you a working familiarity with the material so that the right words come naturally and spontaneously.
4. Enable you to use presentation aids in a smoother, more co-ordinated manner, so that they will strengthen and support, rather than interfere with, the actual presentation.
5. Help you identify and prepare to deal with potential problem areas.

Methods of Practicing

There are four primary methods of practicing a presentation. Any one or a combination of all four can pay tremendous dividends.

1. *Give the presentation aloud to yourself.* Retreat to a room alone with any notes and presentation aids you will be using. Imagine that you are actually making the presentation in front of an audience. Get a feel for the flow of the material. Practice with the aids. Identify any elements that need "polishing."
2. *Videotape your practice session.* The availability of video cameras that can be used in room light has created one of the more desirable ways to practice an important presentation and, during replay, really concentrate on critical elements. Set up the camera in the area where the audience would be. Adjust the lens to isolate the desired field of vision. Again, imagining that you are actually making the presentation, videorecord the en-

tire practice session, including any use of aids. You may wish to critique the recording immediately or view it at a later time. Remember that you are your own best critic! As you watch and listen from the audience's viewpoint, you may discover some things that need additional attention. You may also wish to have some coworkers critique the videotape along with you.

3. *Audiotape your practice session.* Although it does not reproduce the full effect of a presentation, audiotaping does provide you with an opportunity to hear how you sound and to determine whether your ideas are coming across as you desire. Again, you may wish to listen immediately afterward, either alone or with coworkers, or replay the audiotape later.

4. *Give a dry-run.* Have some knowledgeable coworkers, friends, or perhaps even some representative members of the intended audience sit in on your practice session. Although frequently more difficult than the actual presentation, this kind of rehearsal is probably the most effective way to try out your techniques, make sure your ideas are getting across as you want them to, and learn how to field questions on the subject. It is much better to flub a rehearsal like this than to fail at the real thing! *Be certain to provide your dry-run audience with enough background information so that they can gear themselves to react with the knowledge, interest, and attitudes of the intended audience.* You may wish to have your dry-run audience fill out the Presentation Evaluation Guide, shown at the end of this section.

Evaluation and Modification

Your practice session may go smoothly, thereby confirming that your planning and preparation were effective. Then again, you may identify any number of elements that need additional work or even major modification. The key to success with practice is to make whatever changes are called for as a result of your critique. The Presentation Evaluation Guide helps you keep tabs on the many elements of your presentation. Use it to make notes on your practice session and/or to have coworkers evaluate it.

"Practice makes perfect" is a tired cliché and rather ridiculous in this context because you are unlikely ever to give a letter-perfect presentation. On the other hand, "no practice spells disaster" is a realistic statement. The best-prepared presentation in the world can fail to achieve its objectives if it is not presented effectively. Remember, *the preparation process is not complete until you have actually rehearsed your presentation in a practice session!*

Practice Exercises

1. Practice your presentation on the topic you selected earlier, using as many of the suggested practice methods as possible.
2. Have coworkers critique your performance during a dry-run practice session, using the Presentation Evaluation Guide. (Or use the guide yourself to critique a recorded practice session.)
3. Make adjustments in your final presentation based on the results of your practice session(s).

Presentation Evaluation Guide

Topic: _____

Presenter: _____

Evaluator: _____

Content

INTRODUCTION

1. How well did the introduction generate interest in the presentation?

   Outstanding _____   Good _____   Fair _____   Weak _____

2. Was the purpose of the presentation made clear?

   Yes _____   Somewhat _____   No _____   Not sure _____

   Comments: _____

   _____

   _____

   _____

   _____

BODY

1. Did the main ideas come through clearly?

   Yes _____   Somewhat _____   No _____   Not sure _____

2. Were the supporting factual information and any accompanying illustrations:

   Interesting?   Yes _____   Somewhat _____   No _____

   Varied?   Yes _____   Somewhat _____   No _____

   Directly related? Yes _____   Somewhat _____   No _____

3. Was the presentation appropriate for the intended audience?

   Yes _____   Reasonably so _____   No _____   Not sure _____

   Comments: _____

   _____

   _____

   _____

CONCLUSION

1. Did the conclusion summarize the main ideas and purposes?

   Yes _____   Somewhat _____   No _____   Not sure _____

2. How effective was the conclusion in encouraging action, belief, and/or understanding?

   Outstanding _____   Good _____   Fair _____   Weak _____

   Comments: _____

   _____

   _____

   _____

   _____

GENERAL

1. How would you rate the content?

   Outstanding _____   Good _____   Fair _____   Weak _____

2. Do you believe that the presentation objectives are likely to be achieved?

   Yes _____   Probably _____   No _____   Not sure _____

   Comments: _____

   _____

   _____

   _____

   _____

Delivery

PRESENTATION AIDS

1. Were the presentation aids suited to the topic and to the audience?

   Yes _____   Reasonably so _____   No _____

2. Were they visible to everyone and easy to follow?

   Yes _____   Reasonably so _____   No _____

3. How effective was the use of presentation aids?

   Outstanding _____   Good _____   Fair _____   Weak _____

   Comments: _____

   _____

   _____

   _____

   _____

PLATFORM TECHNIQUES

1. Poise: Was the presenter in control of the situation?

   Yes _____   Reasonably so _____   No _____

2. Were posture and movements appropriate?

   Yes _____   Reasonably so _____   No _____

3. Were gestures effective?

   Good _____   Fair _____   Overdone _____   Ineffective _____

4. Was the presenter's relationship with the audience effective (for example, eye contact)?

   Outstanding _____   Good _____   Fair _____   Weak _____

   Comments: _____

   _____

   _____

   _____

   _____

VOCAL TECHNIQUES (check all that apply)

1. How were pitch and voice quality?

   Good _____   Too high _____   Too low _____

   Monotonous _____   Harsh _____   Nasal _____

2. How about rate and intensity?

   Good _____   Too fast _____   Too slow _____   Too loud _____

   Too soft _____   Monotonous _____

3. Did the presenter speak clearly and distinctly?

   Yes _____   Reasonably so _____   No _____

Comments: _____

_____

_____

_____

_____

GENERAL

1. How would you rate the overall presentation?

   Outstanding _____   Good _____   Fair _____   Weak _____

2. Make any additional comments you feel would be helpful. _____

_____

_____

_____

_____

_____

_____

_____

_____

_____

_____

_____

SUMMARY

Any presentation you decide is worth making clearly deserves your best effort to ensure that the intended message is effectively communicated. Only in this way can you achieve a good return on the expenditure of time and money by all involved. The key to effective communication is thorough, systematic preparation.

You start by identifying your objectives for the presentation. If you don't know where you're headed, you could be unpleasantly surprised at where you end up.

A detailed analysis of your audience will disclose a variety of needs, interests, likes, and dislikes. Keeping this valuable information uppermost in mind during the preparation process can enable you to design a tailor-made presentation that will have the greatest likelihood of communicating your message to *this* audience.

The Preliminary Plan is used to identify and document all relevant factors (objectives, data on the audience, main ideas, and supporting factual information). It serves both to keep your own efforts focused and to guide the efforts of anyone who may be assisting you with the presentation.

Selecting appropriate resource materials shouldn't be difficult, assuming this is done in light of the information gathered during the first three steps of the preparation process. Your main task will be to assess material for relevance, establish the degree of detail that will be necessary, and determine the best way to present the resource materials in order to achieve your objectives.

Once the groundwork has been laid, the time has come to organize your material for optimum delivery, breaking it down into an introduction that will not only outline the purpose of the presentation but also sell the audience on your own credibility and the value of listening to what you have to say, a body that expounds on the presentation's main ideas in an appropriate degree of detail, and a conclusion that sends your audience away with a vivid message that you would like them to remember.

Now that you *think* your presentation is all set, it's time for a practice session. This is your opportunity to identify—and remedy—any flaws and to incorporate worthwhile improvements that come to mind in light of your performance during the practice session.

3

Developing and Using Effective Presentation Aids

In the previous chapter, we presented a detailed, systematic approach for performing the considerable work that's required to properly prepare a presentation's content. Having expended all that effort, you might be tempted to believe that your meticulously crafted presentation should now speak for itself. Unfortunately, that's seldom the case.

Of course, if you just happen to be a dynamic speaker, and your message just happens to be brief and compelling, you might be able to get by without supporting visuals and other illustrative material (collectively referred to as *presentation aids*). But if you're like most presenters, you will find the success of your presentation greatly enhanced by skillfully employing visual elements to support your oral message.

Throughout our lives, we are bombarded by visual stimuli. It is therefore only natural to employ visual materials in a presentation. Visuals add an extra dimension of meaning and understanding to your message. They help overcome the limitations of words, simplify complex concepts, and stimulate the brain to remember facts and ideas. Visuals also serve to create points of reference and interest for your audience, help reinforce key concepts, and guide you through the presentation.

And just how greatly can your success be enhanced by supporting your message visually? It has been estimated that people retain about 30 percent of what they are told and about 20 percent of what they see. But that retention rate rises dramatically—to around 50 percent—

for something that is both heard *and* seen. (Even higher retention rates can be attained if people who have heard and seen a message are then given an opportunity to respond to it meaningfully and perform some relevant activity in connection with it.) In short, by engaging your audience on two levels of sensory awareness (seeing and hearing) rather than just one, you increase their involvement and comprehension, thereby increasing your presentation's impact. This makes your audience much more likely to both retain your message and act on it.

This chapter will help you select, prepare, and use presentation aids so as to create this kind of impact. The first major section offers a broad overview of how to translate presentation content into aids that will provide maximum support for your message. The second major section discusses how to use a number of the most common media currently available for displaying your presentation aids, from chalkboards to computers. This section is followed by several practice exercises designed to give you hands-on experience in designing presentation aids and spotting likely sources for visuals. The chapter's highlights are then reviewed in the Guidelines for the Effective Use of Presentation Aids, which can serve as a ready reference for the presenter. The Supplement to Chapter 3 "Effective Aids to Understanding" briefly summarizes the nature and purpose of almost every conceivable type of presentation aid. As a tool for assessing your available presentation-aid options, it can help you quickly and effectively decide on the best way to illustrate your main points.

TRANSLATING YOUR MESSAGE
INTO VISUALS

In any presentation, there are only a few key ideas (the ones you outlined in your Preliminary Plan) that must be communicated to the audience. By concentrating only on key ideas, you can often conjure up mental images to capture these ideas and identify concrete results you want to achieve with a given visual aid.

For example, let's say that your firm is subject to the jurisdiction of a federal regulatory agency. That agency announced several months ago that a series of new regulations will take effect on January 1 of next year. They affect one of your major operating departments, which has eleven product-processing lines. Once the regulations take effect, you will have two years in which to bring that department's operations into compliance. This will primarily involve purchasing updated equipment, retraining the personnel assigned to each line in the use of the new equipment, implementing revised quality-control procedures, and utilizing a very different record-keeping system as soon as the new equipment for each line is in operation.

You are briefing upper management on the findings of the company task force you headed that was assigned to study the impact of the new regulations and develop a proposal for bringing the firm into compliance with them. One of your main points is to show how your task force's proposed time frame will bring the eleven product-processing lines into compliance during the two-year phase-in period.

There are many ways you could illustrate that time frame. You might use a line graph, with the horizontal axis indicating months and the vertical axis indicating number of product-processing lines in compliance (figure 3-1), or you might use a multibar graph, with a series of bars of varying lengths representing the number of product-processing lines brought into compliance as of the first day of each month during the phase-in period (figure 3-2).

In choosing among equally valid visual options for helping you make a particular point (such as the options shown in these two figures), try to be guided by how you expect your audience to react to each type of aid.

Now that you've seen how to translate one key point into a visual, we'll present a technique, called storyboarding, for systematically analyzing your entire message and translating its content into presentation aids. Then, in the remainder of this section, we'll discuss some critical

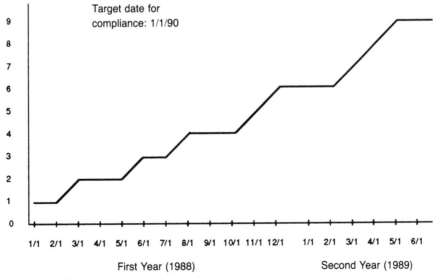

FIGURE 3-1

One Option for Illustrating Proposed Progress toward Compliance with New Regulations

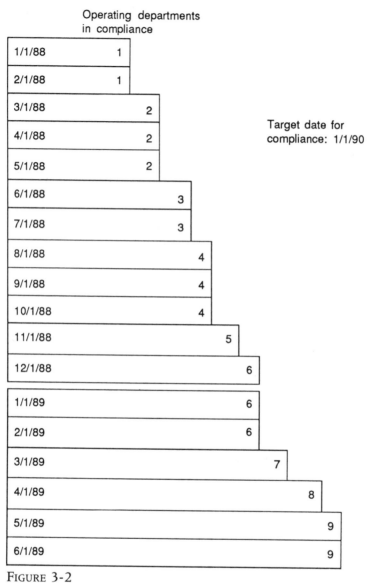

Operating departments in compliance

Date	In compliance
1/1/88	1
2/1/88	1
3/1/88	2
4/1/88	2
5/1/88	2
6/1/88	3
7/1/88	3
8/1/88	4
9/1/88	4
10/1/88	4
11/1/88	5
12/1/88	6
1/1/89	6
2/1/89	6
3/1/89	7
4/1/89	8
5/1/89	9
6/1/89	9

Target date for compliance: 1/1/90

FIGURE 3-2

Another Option for Illustrating Proposed Progress toward Compliance with New Regulations points in every presentation that require illustration, provide tips on where you can obtain suitable artwork to create your visuals, and present general guidelines for the selection and preparation of presentation aids that will most effectively support your message.

What Is Storyboarding?

Obviously, illustrating every point you make during a presentation would be boring and counterproductive. So, how do you decide what to illustrate? A helpful technique for highlighting major points and identifying ways to translate them into pictures is called storyboarding. It involves listing key concepts and ideas on the left-hand side of a page and translating each concept into a visual in a corresponding box on the opposite side of the page (figure 3-3). Some presenters prefer using 3" × 5" index cards for storyboarding (figure 3-4)—writing a point at the bottom of each card and then rough-sketching or describing an image on the top portion of the card or attaching a photo or clipping to represent the key idea. (A blank storyboard form can be found in the appendix.)

Storyboarding helps you to generate, collect, and organize visual ideas—and coordinate them with your content in a way that achieves maximum impact. It also allows you to preview the flow of your entire presentation and get feedback from colleagues. A completed storyboard is extremely helpful, as well, to whoever produces the presentation aids, be that your secretary, an internal or contract graphic arts specialist or yourself.

Critical Points to Illustrate

In every presentation, regardless of its specific content, there are certain key portions of your message that require illustration, particularly the introduction and the conclusion as well as the main ideas in the body.

You may want to start your presentation with a title visual or something else that commands attention. The moment you present a visual stimulus, your audience will respond to it. Therefore, make sure that your opening visual focuses on what you want the audience to think about first—namely, the purpose or concept of the presentation. You can use an appropriate graphic design or picture to personalize your title visual to a specific audience. For example, you might kick off a report to southeastern account executives on sales projections for the coming year with a visual that features an outline of the geographic

Key Concepts	**Visual Representations**

Welcome. Announce new profit levels. Establish need for further growth.

Photo of hdgtrs. bldg. with words "SUCCESS CONTINUES" overprinted

How company plans to engineer growth.

(Get graphic from art dept.)

Percentage of revenue budgeted for new campaign.

(pie chart)

New sales aids to be distributed.

 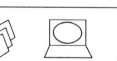

(photo of new aids)

FIGURE 3-3
Sample Storyboard

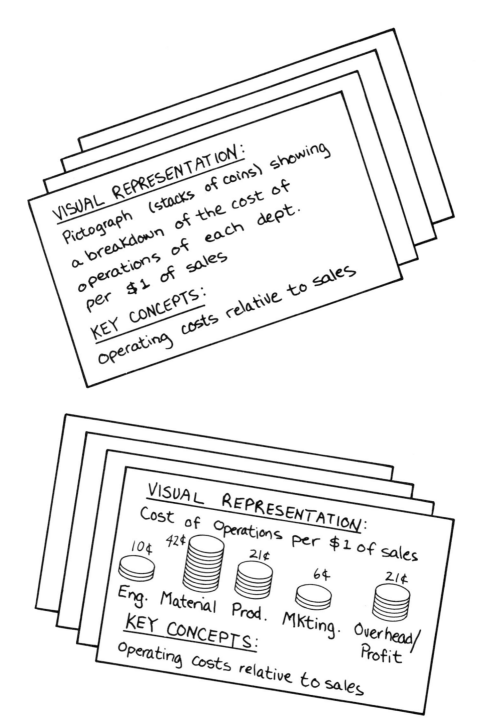

FIGURE 3-4
Two Ways to Use 3″ × 5″ Cards for Storyboarding

**YOUR CHALLENGE
IN THE COMING YEAR**

FIGURE 3-5
Sample Opening Visual

territory overprinted with a suitable theme-setting message (figure 3-5).

Your second visual usually presents the agenda or goals of the presentation. It lets the audience know what to expect and shows that you have carefully organized and prepared what you are going to say. Again, use key words or phrases. Fill in detail orally or provide the audience with a handout. Don't clutter the visual. Keep it simple.

From this point, proceed into the body of your presentation. And don't forget to close with a conclusion or summary graphic that leaves your audience with a vivid image of the message you want them to retain.

Where Do You Get Your Visuals?

Many presenters keep an art file that includes shapes, outlines, drawings, and cartoon characters clipped from magazines, newspapers, company publications, and other published sources. Brainstorming freely with colleagues and customers can help you identify additional sources. There are also commercial packages available containing hundreds of visuals that can be used as is, combined, and adapted in many ways to create a wide variety of interesting presentation aids. If a graphic artist is available, have him or her draw some stock visuals, or use illustrations from past presentations in new combinations (with appropriate modifications, of course). Keep photographs or slides of items or events you may wish to use in the future, as well.

The key to the availability of raw material for visuals is to think of the content of your most frequent presentation topics from a visual standpoint and keep your eyes open constantly for suitable sources of line art and photos.

General Guidelines for Selecting and Designing Presentation Aids

The rules in the accompanying box and the following basic principles should be kept in mind when choosing and preparing presentation aids.

An Effective Visual Aid Must . . .

▲ Present an idea better than speech alone.

▲ Represent a key concept.

▲ Support only one major idea (even though it may present a lot of information about that idea).

▲ Emphasize pictures or graphics rather than words wherever possible.

▲ Restrict the use of text to a maximum of six words per line and ten lines per visual, with short phrases or key words employed rather than complete sentences.

▲ Use color or contrast only to highlight important points.

▲ Represent facts accurately.

▲ Be carefully made: neat, clear, uncluttered.

▲ Have impact.

GROSS SALES			
1976	$3,410,740	1981	$ 4,402,225
1977	$3,676,490	1982	$ 3,703,291
1978	$4,247,613	1983	$ 6,212,116
1979	$4,750,255	1984	$ 7,864,950
1980	$5,227,671	1985	$11,338,876

FIGURE 3-6

A Typical but Not Particularly Effective Way of Presenting Facts and Figures

CLARITY

Is the aid simple and easy to read? Does it clarify and give substance to your ideas through pictures? Or does the aid try to do too much at once? Some presenters feel that the more visual information provided, the better, so they design their entire presentation accordingly. Yet visuals that are overcrowded with words, numbers, and/or images only muddle concepts.

For example, a history of gross sales over the past ten years could be shown in terms of columns of figures (figure 3-6). But while the information it displays is accurate, this visual fails to communicate a clear message. Rather, it comes across as an unfocused collection of facts and figures. The same information, when shown as a bar graph (figure 3-7), vividly depicts the relationship of the numbers to each other—and does so in a way that would permit gross sales to be readily compared with similar bar graphs showing factors (say, demographic changes) that may have had an impact on sales during this period.

APPROPRIATENESS

Is the presentation aid appropriate to the audience? To the occasion? To the speaker? Does it fit your message? A presentation aid should be designed or selected to illustrate some particular point you are trying to communicate. Don't modify your points to make them fit the aid you happen to have available. Instead, determine the kind of reaction you want from your audience and the kind of image you want created of you and of your presentation. Then make certain your presentation aids project this image.

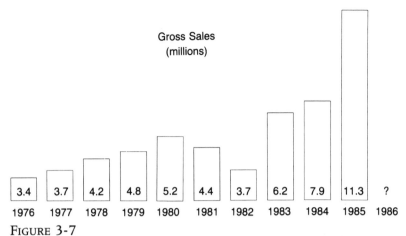

Gross Sales
(millions)

| 3.4 | 3.7 | 4.2 | 4.8 | 5.2 | 4.4 | 3.7 | 6.2 | 7.9 | 11.3 | ? |
| 1976 | 1977 | 1978 | 1979 | 1980 | 1981 | 1982 | 1983 | 1984 | 1985 | 1986 |

FIGURE 3-7

A More Effective Alternative for Visually Displaying Facts and Figures

ARRESTING QUALITY

Is your aid an imaginative tool that focuses attention on the topic under consideration, or is it a gimmick that draws attention only to itself? A group of nude figures might be a novel way of calling attention to a group of significant financial figures. But which of the two sets of figures will your audience remember after the presentation? An effective presentation aid should emphasize the *subject matter* you want the audience to recall. It does not necessarily have to be flashy. The careful use of color, boldface lettering, underlining, arrows, and other such symbols and techniques can highlight those factors you wish to accentuate.

QUANTITY

Are your presentation aids so numerous as to risk overstimulating your audience? A good rule of thumb is that a presentation should normally contain no more than one visual for each two minutes of presentation time. That means a maximum of seven or eight visuals for a fifteen-minute presentation.

Although numerous exceptions could be cited, the at-least-two-minutes-per-visual rule is still valid. It takes twenty to thirty seconds just for your audience to focus on the content of a visual, especially if it is in any way complex or is a type of chart or diagram they aren't used to seeing. Also, if the point is important enough to be illustrated in the first place, it is important enough to justify your spending sufficient time to let the audience absorb the facts.

Many presentations include too many visuals, and timing their display will give you a standard by which to control their number. Naturally, your specific topic, your audience, and the nature of your presentation may call for a deviation from this rule. But remember that the one-to-two ratio is a good starting point, and if you do find yourself tempted to make an exception, check to be certain that you really *need* all the visuals.

AUXILIARY NATURE

The most important thing to keep in mind is that an aid should be just what its name implies: it should *support* the presentation, not be the center of attention. The content of a good presentation should be able to stand on its own *without any aids*. The aids should strengthen what you are saying but should require your interpretation. Otherwise, you become merely a robot that operates the aids.

PRESENTATION MEDIA

Now that you've selected or designed the type of presentation aid you want (a pie chart, for instance), a critical element in planning involves choosing the means to display this visual. Will you draw your pie chart on a chalkboard, show it with an overhead projector, present a slide of it, print it as a handout, or display it on a computer screen?

In addition to the more traditional means, recent technological advances have given us a number of new options. Your choice will depend on such factors as the type of presentation you are making (persuasive, explanatory, instructional, oral report), the audience, time and cost limitations, and the nature of the medium. Although there is no magic formula for instantly and accurately deciding which medium to use, some knowledge about their capabilities and limitations may help you make the decision.

Our experience indicates that about 60 percent of all business and technical presentations involve the use of overhead projection; another 20 percent are supported by slides; and the remaining 20 percent use flipcharts, handouts, props of one sort or another, computers, video, recordings, or combinations of all of the above.

In the remainder of this section, the discussion will focus on overhead projection, slides, and video, followed by a brief overview of a number of other, less mechanically complex presentation aids (such as the flipchart, handouts, and workbooks). We will then examine the critical role of practice in ensuring that your use of presentation aids is an asset rather than a liability to your presentation. But first, we'll touch on a few fundamental considerations that apply regardless of the particular medium being used.

General Guidelines for Using Presentation Media

Despite your careful attention to the selection or design of a presentation aid, its effectiveness could be seriously undermined if the means you employ to present it is not satisfactorily seen or heard, cannot be easily handled, or is not readily accessible or available.

VISIBILITY OR AUDIBILITY

Can your presentation aid be seen or heard satisfactorily by everyone? In other words, have you taken into consideration the size of the room, the number of people, proximity to a source of distracting noise, the seating arrangement, visual obstacles, lighting, and so forth? An aid that cannot be properly seen or heard is worse than no aid at all: it will irritate and distract your audience.

The rule regarding audibility of presentation aids applies equally to the presenter's own voice. While you will ordinarily speak somewhat louder than normal conversational volume in delivering a presentation, it is particularly important that you *project* during the use of presentation aids. The listeners' attention will at that point be divided between you and the presentation aid, so you will have to speak up to hold their attention. Furthermore, if you are showing slides or a motion picture, you should be aware that greater volume is required to hold attention in a darkened room.

EASE OF OPERATION

After suitable rehearsal, can the aid be handled easily? If the aid does not lend itself to smooth presentation, your difficulties in working with it will distract your audience and detract from the effectiveness of your message. (See the guidelines in the accompanying box.)

To Simplify Your Handling of
Presentation Media . . .

▲ Set up and focus any equipment before the presentation. Minimize the mechanics of operation that you must deal with during the session.

▲ Arrange components (transparencies, parts of a model, slides) in proper sequence beforehand and have them handy to wherever they will be needed.

▲ Designate someone to help with lighting, if the lights must be turned on and off.

▲ Leave presentation aids, particularly equipment, intact until after the audience is dismissed. Don't try to rewind a cassette, rearrange slides, remove flipchart pages posted on the wall, and so on while the session is still underway.

ACCESSIBILITY OR AVAILABILITY

Will the aid or equipment for showing the aid be available to you at the presentation site? Can it be made available at a cost (in terms of money, time, or convenience) that will not exceed its relative value to the presentation? Can it be stored in the order in which it is to be utilized but kept out of sight of the audience before and after use? A model of a particular piece of machinery, for example, can be an effective aid, but it will distract the audience if it is left in view throughout the presentation. If such aids cannot be placed completely out of sight, you can keep them covered when they are not actually in use.

Overhead Projection

Because an overhead projector is easy to use, and its transparencies are inexpensive and uncomplicated to produce, it is a natural for the fast-paced world of presentations.

Projectors come in all sizes and shapes, from briefcase portables to console models. Their capabilities are similar, although images projected by the lightweight portables often lack the clarity of those projected by the larger models.

Most standard lenses are designed for use in situations where the presenter is in the front of the room facing the audience, with a projection screen behind him or her. The throw of the lens is twelve to fifteen feet. Lenses that allow you to adapt the overhead projector for use with large groups or very small corporate boardrooms are also available.

One difficulty with overhead projection is keystoning, or the distortion of the image due to the proximity of the projector to the screen. This can usually be overcome by adjusting the angle of the screen. Burnout of projector bulbs is another common problem, which has been solved by the availability of projectors with dual bulbs. When one bulb burns out (smack in the middle of your important presentation, as is so often the case), you simply flip a lever, activating the spare bulb, and continue.

An overhead projector is not only simple to operate but also offers you maximum control and flexibility as a presenter. It allows you to maintain eye contact with the audience while still being able to read what is being projected on the screen behind you. The overhead projector is also designed to be used in ambient light conditions. Since there is no need to darken the room, the rapport you've built with an audience won't vanish in the shadows, nor will anyone taking notes jeopardize their comprehension by struggling to write in the dark.

HOW DO YOU WORK WITH AN OVERHEAD PROJECTOR?

When you place your transparency on the projector table, the projected image will become the center of attention. Control that attention by leaving the projector off until you are ready for the transparency to be seen. As you present succeeding transparencies, learn to transfer smoothly from one to another. And when you have finished with the transparencies for the moment, you should turn the projector off and draw attention back to you as you summarize major concepts.

In using a transparency to outline a series of points, it is often helpful to avoid disclosing them all at once, because the audience will have a tendency to "jump ahead of you." If they are scanning through the later points, they will not be giving their full attention to your discussion of the earlier points. To prevent this distraction, you can mask specific items on a transparency by sliding a piece of paper between the transparency and the projector table and then moving the paper to reveal each point in turn, as you are ready. Overlays can create the same effect. A basic title can be shown on a transparency mounted in a frame. Succeeding items can be added through the use of overlying strips taped to the transparency's frame (figure 3-8), which are then flipped down to build a complete image, one portion at a time.

Besides serving as a protective perimeter for transparencies and a place to tape overlays, the frames themselves have many uses. They provide a handy place for writing short notes to jog your memory. When preparing transparencies for a presentation, you can also use the frames to mark those that are essential to conveying your message. Then, if your presentation must be shortened, you can do so without losing rhythm. Using the frames to number the transparencies is a must. Not only does numbering help you to remember the transparencies' sequence, but it will also prove invaluable if you ever drop them.

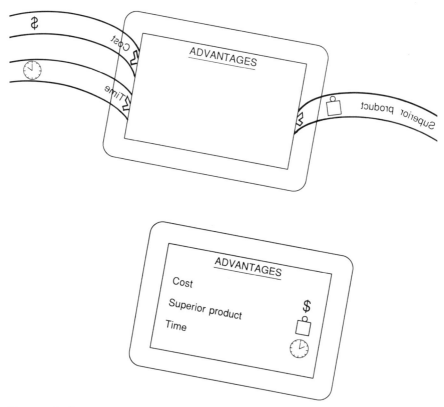

FIGURE 3-8
Transparency Overlays

The overhead projector gets its fair share of misuse. Most often, this involves employing overhead projection to display a long section of printed text that the presenter reads. Effective transparencies show only key words or phrases and represent concepts pictorially whenever possible.

Another typical misuse of the overhead projector involves the presenter's losing eye contact with the audience. In working with a transparency, take special care not to stand next to the projector and stare down at the transparency for extended periods of time, leaving the audience with a view of the top of your head. And if you need to emphasize something, don't turn away from the audience and look at the screen. Instead, remain facing the audience and point to the item on the transparency itself with a pen or pointer. An alternative is to use a pointer on the screen, using the hand closest to the screen to hold the pointer, so as not to block your audience's view. Remember: the audience, not the equipment, should get your eye contact.

HOW DO YOU CREATE TRANSPARENCIES?

Transparencies are prepared in a number of ways. The simplest is to write/draw/trace on a blank transparency with special marking pens. Both pens whose markings are permanent and pens whose markings can be wiped off (either with a cloth dampened in water or with a cloth dampened in alcohol) are available. (Know which type you are using!)

Another common preparation method is to use a transparency-making machine, which employs a thermal carbon-lifting process to burn images onto a sheet of acetate. Usually, a master is prepared by arranging art, illustrations, borders, figures, words, and so on, on a piece of paper exactly as you want them to appear. A photocopy of the master is then processed through the transparency-making machine. Some office photocopiers may also be used to prepare transparencies. Color can be added with transparency marking pens. Another way to create colored images is to use special tinted plastic sheets that can be cut to the appropriate size and shape. Once the peel-off backing is removed to expose an adhesive coating, these very thin sheets can easily be affixed to the transparency's surface, providing colorful highlights but without adding greatly to the thickness of the transparency.

Probably the most startling impact on transparency production has been caused by the introduction of the computer. Graphics software and high-resolution color printers now make the production of quality transparencies quite simple. Though transparencies created with the assistance of a computer are more expensive than the hand-made variety, the results sought from a particular presentation may justify the higher cost per transparency. The same design and presentation techniques are applicable; only the method of production changes. Each computer system has its own methodology and often requires the assistance of a professional.

Large, clearly defined letters and numbers and simple illustrations should be the rule in preparing transparencies. Typewriter-sized letters, even those produced by an Orator-type element, are not adequate. Letters and figures should be at least a quarter-inch high. Press-on letters or those produced using a lettering machine should be at least "ten point" in size. Hand lettering should be meticulously printed.

Transparencies are easy and fun to make. In fact, you are encouraged to experiment with producing them yourself or having an assistant develop them. Their preparation is a refreshing change of pace from routine work and allows the creator a real sense of personal satisfaction. For those who want their transparencies professionally produced, the storyboard (discussed earlier in the chapter) will be the road map that the graphic artist uses in designing them.

What about Slides?

The ready availability of cameras and slide film and the rapid emergence of computer-generated graphics have made slides a favorite among presenters who want to add color, resolution, and versatility to their presentations. Slides may be used to show everything from detailed line drawings to actual photographs of products or people. Their crispness and professional image tell the audience that your presentation has been carefully planned and well thought out.

Slide projectors vary mainly in their array of special features, but the most common models all accept standard 35 mm (2" × 2") slides. If you are purchasing a projector or are indicating specifications for equipment needed for a presentation, ask for a projector that has remote-control capability and a zoom lens (102 mm to 152 mm, f/ 3.5). You will be able to control the flow of slides from your position at the front of the room, and you can adjust the size of the projected image to match the size of the room and the audience.

Most slide projectors have a round tray on the top of the projector that holds up to 120 slides. On many projectors, the tray rotates in a clockwise or counterclockwise fashion, with the slides being displayed in the sequence in which they are placed in the tray. This arrangement can limit your flexibility in sequencing. The new generation of random-access slide projectors helps overcome this limitation, but their use also requires careful planning.

Some slide projectors can be hooked up to a synchronized audiotape machine, which plays a prerecorded tape containing signals that cause the projector to change slides automatically. (This is the proverbial "canned" presentation, which could be appropriate, depending on the results you desire.)

Designing slides involves the same rules of thumb that apply to other visual aids. You should use creative pictures in lieu of raw text or raw numbers whenever possible. When text is essential, fill no more than half the viewing area. Limit each slide to one visual idea. It is better to use several simple slides than one complicated one. Vary the design of each slide to avoid visual boredom or use clearly distinct titles so that the audience will immediately notice a change in content. An overview followed by a close-up photograph can be used to present important details in context. Use color for contrast and to highlight important points.

Layout of copy for slides is more complex than for overhead transparencies, resulting in a slightly higher cost for slides and generally requiring professional graphic art support. When appropriate, it is quite simple to actually take photographs using slide film or to copy visuals that have been developed for other purposes (annual reports, sales brochures, and so on). These can be interspersed with stock visuals—slides of symbols, characters, and figures that are available commercially or that can be produced for your special needs to provide presentation support.

Many organizations and professional slide producers generate slides using in-house computer graphics systems. Computer graphics capabilities include multiple fonts of type, extremely versatile layout and design functions, and an infinite array of colors. Company logos, photos, product visuals, and so forth can often be digitized and stored in the computer's memory for recall during slide development. Light pens, mouses, and touch-screen features can be used to augment the keyboard in accessing the many images capable of being generated by the computer. A finished visual is displayed on a special high-resolution video screen and then photographed with a camera hooked up to the system. Although capable of being operated by a novice, these systems are usually complicated and expensive enough to require the services of a graphics or computer specialist to handle design and production.

In view of the cost involved, some form of storyboard is practically a must in planning a presentation that will require slide support.

HOW DO YOU USE A SLIDE PROJECTOR?

The location of slide projector and screen historically has presented problems. Usually in a meeting room, the screen is centered in the front. The presenter must then stand off to the side, and the screen is literally the center of attention. In large banquet halls, the presenter can often be found on one side of the room and the screen on the other. Which way should the audience look? In either of these cases, the screen/projector setup makes the presentation aid the focus of audience attention rather than an adjunct to your delivery.

In a meeting room or banquet hall, the ideal setup is for the presenter to be in the front of the room at the center with the screen to the audience's right at about a 45° angle to the audience. The projector should be located on an elevated stand or cart at the side of the room, toward the rear, on the audience's left (figure 3-9).

The location of the screen is to the audience's right because the audience, having been trained to read from left to right, will likewise be visually conditioned to view information from left to right. You can therefore take advantage of the audience's natural tendency to characterize *you* as the focal point of the presentation, sweeping their eyes to the right to fix in their minds a concept illustrated by a visual and then returning automatically to you.

The size of the projected image should be adjusted to match the size of the audience. Measure the distance from the screen to the farthest member of the audience. The image width should be, at a minimum, one-sixth of that distance. That means, for example, that if the farthest viewer will be 36 feet away, the width of the projected image should be at least 6 feet. Don't let the size of the image overpower the message, however. Remember that your visual is an *aid* to your presentation. Also, don't permit the audience to sit closer than two image widths from the screen (12 feet in the example).

In certain facilities, the presenter may have no control over screen placement or may choose to use some other option for screen placement (such as rear-screen projection). In these cases, consult an audiovisual specialist for guidance or use good common sense.

EFFECTIVE APPROACHES FOR ARRANGING SLIDES

Naturally, the slides should be arranged in such a way as to complement the type of presentation you are making (persuasive, explanatory, instructional, or oral report). There are four commonly used

80

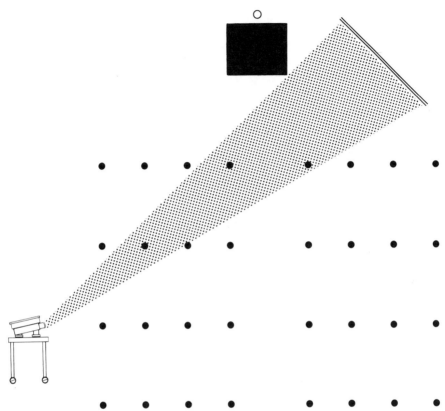

FIGURE 3-9
Proper Placement of Slide Projector and Screen for Use in a Presentation

arrangements that can help clarify your presentation: progressive disclosure, sequencing, serial, and question/answer.

Progressive disclosure is used most often when text and numbers are displayed. This approach allows the audience to see one thought at a time, followed by another and another (successive lines of text are a good example). To create this slide sequence, the complete visual is laid out and then a series of slides is made, starting with one showing the last item masked out, followed by another showing the last two items masked out, and so on until the last slide shows everything masked out except the title and the first item on the list.

These slides are projected in reverse order from the way they were created—that is, starting with the slide showing only the title and the first item on the list, followed by the slide showing the title and the first item on the list plus the second item, and so on until the complete

PROGRESSIVE DISCLOSURE

- **Helps audience keep up with presenter**

PROGRESSIVE DISCLOSURE

- Helps audience keep up with presenter
- **Reveals key ideas one at a time**

PROGRESSIVE DISCLOSURE

- Helps audience keep up with presenter
- Reveals key ideas one at a time
- **Generates a list of ideas for review**

PROGRESSIVE DISCLOSURE

- Helps audience keep up with presenter
- Reveals key ideas one at a time
- Generates a list of ideas for review
- **Helps to summarize at the end**

Figure 3–10
Progressive Disclosure

visual is displayed (figure 3-10). This way, as each succeeding slide is projected, the audience's attention focuses on the new thought that is revealed. And since the audience is prevented from reading ahead of the presenter, their attention remains on the point under consideration. Changing the visual as each new idea is presented also tends to stimulate the audience's interest.

Sequencing is similar to progressive disclosure but involves pictures rather than text. A sequence of related visuals developing a single concept is projected in building-block fashion. An example is the stages in the manufacture of a product, starting with design and ending with the finished item. Another is a pie chart with each slice projected in turn for individual consideration and then all the slices joined together to represent the whole. Sequencing can strengthen the audience's comprehension of an idea.

Serial slides, which involve comparing several related items by showing them in a series, can also be a beneficial arrangement for a presentation. The evolution of a certain model of car over the years is a good example, as are before, during, and after shots of a construction project. Serial presentation also engages the full power of the audience's memory.

Question/answer slides depict a direct relationship between the statement of a problem and its solution. A slide showing a question written out or a representation of some goal can focus the audience's attention on your point while you expound on it orally. A subsequent slide can show the correct or desired answer. This technique is often used in interactive presentations, where the question slide serves to generate audience involvement and the answer slide summarizes the discussion.

Video: An Option You Should Consider

The fact is, television is here to stay, and it is coming to have a profound impact on business and technical presentations. It is therefore of increasing importance for presenters to learn how they can meaningfully utilize the tool of nonbroadcast television (that is, video) in a variety of situations to deliver a well-planned, high-quality message.

Videotaping involves an electromechanical procedure by which both a picture and sound are recorded on special magnetic tape, much as a conventional tape recorder records just sound. Videotape itself comes in several different sizes, but commercially distributed tapes are usually packaged in cassettes, using either ¾", ½", or 8 mm videotape. A playback system typically includes a video playback unit, a television monitor, and connecting cables. A generation of small, portable videocassette recorders (VCRs), which use the newer ½" or 8mm videocassettes and which are priced well within the reach of businesses and many individuals, has made this technology increasingly available to those who prepare presentations.

The practical applications of video are limited only by the presenter's imagination and willingness to experiment with this versatile technique. A few examples:

▲ When the audience must see the performance of complex or intricate maneuvers involving specialized equipment or minute system components, a strategically placed video camera can capture the action and multiple monitors or large-screen projection can provide everyone with a front-row seat.

▲ Presentations that must be made to a large, geographically diverse audience can be beamed via satellite to any place on earth equipped to receive the signals—a process called videoconferencing.

▲ Testimony (from an expert, a consumer, a top executive) supporting one or more of your major points can be delivered via video (and preserved for reuse in later presentations), freeing the presenter from the limitations associated with having such testimony delivered live.

▲ When audience members must demonstrate certain skills during the presentation, they can be videotaped using a simple setup. The tape is then played back and critiqued.

WHERE DO YOU GET VIDEOS?

The videos used in presentations come from a variety of sources. These include real-time video, in which the presentation is broadcast live via closed-circuit television; recorded video, in which the audience is shown either a commercially available video that has been rented, leased, or purchased, or a video that has been produced in-house; and computer-generated visuals. If you wish to produce your own videos in-house, this subsection raises some points to keep in mind.

In videotaping, as in a live presentation, almost every type of auxiliary material can be adapted to and incorporated within the framework of the presentation. Charts, slides, overheads, photographs, movies, music, sound effects, dialogue, and special effects utilizing any or all of the foregoing can be effectively combined to ensure that audience interest and comprehension are maximized and the greatest benefit is derived from the presentation.

A televised presentation or segment of a presentation must, by its nature, be more tightly planned and structured than a live presentation. Developers of video sequences, examples, and dramatizations used to augment presentations must always keep in mind that members of the audience have in all probability already seen thousands of video programs on their home television sets. Even though videos prepared for use as presentation aids need not live up to Hollywood standards, certain basic rules for television production still apply. Chief among them is the fact that television is a medium of movement, and movement must be regarded as a vital component in designing video aids. While shots of people standing motionless before a camera and speaking (sometimes referred to as "talking heads") and static shots of text do have a place in video, they should be used judiciously and sparingly.

The same planning and storyboarding techniques appropriate for other visual aids are critical when it comes to video, given the medium's cost. The presenter who opts for a relatively elaborate video program, however, should also get professional help during the production stage.

HOW DO YOU WORK WITH VIDEO?

The presenter's personal involvement in setting up the video equipment will depend on the complexity of the equipment, which, in turn, often depends on the size of the audience.

Many presentations in which video is used are for smaller groups. As a general rule, a group of up to twenty-five people can comfortably view a nineteen-inch video screen, although twenty-five people is a maximum. In such a situation, the presenter will have no trouble operating the necessary equipment.

Equipment setup for large groups is best handled by an audiovisual specialist. One option is for several individual monitors to be hooked together; another option is the use of video projection equipment, in which an image is displayed on a large screen manufactured especially for video. The video projectors themselves can be quite complex and require exacting alignment with the screen. Computer-enhanced and distributed video should also be set up by someone specially trained to do so. This is not to say that you are incapable of performing these tasks yourself. If you've been properly trained and are familiar with the technology, have at it! But don't risk the success of your presentation by attempting something beyond your capabilities.

When Something Goes Wrong with Your Equipment . . .

Seasoned presenters know that Murphy's law applies particularly to presentations. You would therefore be wise to pack yourself an "emergency kit" for both on-site and off-site presentations.

Spare bulbs for whatever projectors will be used (even for the dual-bulb models) head the list, followed closely by a three-pronged electrical adapter. An electrical extension cord packed in the kit has saved many a presentation when the outlet was just a little too far away and no one from maintenance could be located! A remote-control extension cord for a slide projector will enable you to change slides without having to remain within an eight-foot radius of the projector.

Masking tape—an everyday item that you might not immediately think of when stocking an emergency kit—has made things a lot easier for scores of presenters. From covering exposed cables and cords for safety purposes to posting flipchart pages on the wall, masking tape can be a presenter's best friend!

And finally, the telephone number of the in-house audiovisual specialist is a must in the event that *nothing* works. In that worst-case scenario, remember that a good presentation should be a success even without the use of any presentation aids whatsoever!

Other Presentation Aids

Although overhead projection, slides, and/or video are used to support the vast majority of business and technical presentations, there are a number of other aids that can enhance any presentation.

RECORDERS

Occasionally, the use of some sort of recording can be an effective way of presenting material. Recordings of key people who cannot be present may add prestige as well as content value and provide a change of pace for the presentation. Sometimes a recording of an interview or a conference can be used to advantage. In certain cases, a sound effect (such as the sound of machinery in operation or a test being conducted) may add dramatic impact and support your objectives.

A **record player** might be useful if you have a professionally produced record that happens to be relevant to your topic. However, a record player is not a very practical tool if what you need to present is your own material.

An **audiotape recorder** is the easiest, least expensive, and most practical type of recorder for most purposes. In addition to the larger, more standard models, there are many small, inexpensive, high-quality recorders that can either be plugged into an electrical outlet or run on batteries. The audiocassette recorder offers the added advantage of being virtually "operator-proof." This sort of recorder is adequate for most presentation requirements. The audiotape recorder is particularly useful in conjunction with a slide presentation. It can be synchronized with the slides to provide appropriate sound or a running commentary for what is being seen on the screen.

CHALKBOARD

The chalkboard is one of the most useful, most available, and least-expensive types of visual aid equipment you can use in a presentation. Despite its familiarity, it is not an easy aid to work with and requires advance planning and practice if it is to be employed effectively.

The chalkboard offers the presenter a number of advantages:

1. *Flexibility.* It provides plenty of space on which all sorts of material (words, diagrams, sketches) can be displayed and changed relatively easily.
2. *Sense of immediacy.* By writing material on the chalkboard, the presenter gives the audience the feeling that this is the very latest information—even more current than material displayed on charts, which could have been prepared some time ago.
3. *Progressive development.* Starting with an initial step (as in a flowchart) or with a simple base (as in a model diagram), an idea, process, or design can be developed progressively on the board while you explain each step. As with the progressive-disclosure approach used with slides, progressive development focuses the audience's attention on one step at a time and prevents their getting ahead of the presenter and hence becoming distracted.
4. *Feeling of spontaneity and audience involvement.* If you become skilled in the use of a chalkboard, you will be able to request comments, questions, and other input from your audience and record these contributions on the board as you go along. This not only heightens the audience's sense of involvement but shows that you are both willing and able to be immediately responsive to their needs and interests.

5. *Change of pace.* Writing on a chalkboard adds variety to your presentation, helps maintain audience interest, and gives the audience a chance to take notes on key points.

The following guidelines can help you work more effectively with a chalkboard:

1. Write legibly and neatly, occasionally combining longhand and printing for emphasis. Write much larger than usual.
2. Hold the chalk at approximately a 45° angle and apply sufficient pressure to make a fairly heavy line. If the chalk squeaks, break it in half. (A metal chalk holder, available in most stationery stores, will reduce or eliminate squeaking, breaking, and dust.)
3. In order not to distract the audience, keep material that is put on the chalkboard ahead of time covered until you need it; likewise, erase material from the board or cover it when it is no longer needed.
4. Allow enough time for the audience to copy or study the material displayed on the chalkboard, because once it is erased, it is gone for good.
5. While you are writing on the chalkboard, maintain audience contact by keeping up your flow of talk. Insofar as possible, avoid talking directly to the board.
6. Try not to write lengthy material on the chalkboard during the session. When you must write a series of items at once, step back from the board every few seconds to let the audience see the material.
7. Use colored chalk (sparingly) to highlight major points.

FLIPCHART

The flipchart, consisting of a large pad of paper mounted on an easel, offers many of the same advantages as the chalkboard. Simple diagrams, key words, and outlines can be drawn or written on it with broad-tipped marking pens, and responses from the audience can be recorded on it during interactive sessions.

The flipchart also provides several distinct advantages over the chalkboard. There is no need to erase material displayed on flipchart pages. A completed sheet can simply be flipped over to reveal a fresh sheet beneath, or individual sheets can be detached from the pad and mounted on the wall with masking tape (if allowed) for continued reference. In either case, fresh flipchart surface can be freed up without the need to obliterate existing material, as must be done when material on a chalkboard is erased to enable additional material to be displayed. And of course, the problem of chalk dust is eliminated.

Entire flipchart pages can be prepared in advance, or principal talking points can be noted briefly in pencil on flipchart sheets. Unseen by the audience, these notations can help you remember your main points. When preparing flipchart pages in advance, leave a blank sheet in between prepared pages to keep the marker from bleeding through. Once you've finished working with a prepared page and have flipped it over or detached it for display on the wall, the blank sheet also serves to keep the audience from seeing the next prepared page until you're ready to begin discussing it.

HANDOUTS

Handouts are another common and helpful presentation aid. They can provide detailed technical information, guide individual study and practice, and present administrative information such as objectives and evaluation procedures. However, they are effective *only* if careful thought has been given to their preparation and use. Otherwise, they are costly and wasteful, and may even jeopardize the accomplishment of your objectives.

Normally, handouts should be distributed *following* the presentation, unless you envision their use *during* the presentation. For example, a new form that you will be explaining and working actively with in the course of the presentation should be made available to the audience for reference as you project an overhead transparency or slide of it. On the other hand, detailed financial figures documenting the trends you may be highlighting serve only a supplementary function and are best not distributed until afterward, so you don't invite the audience to look at the handout when they should be listening to you.

As you go along, you can build audience interest by indicating something new, different, or particularly useful to be found in the handouts that will be distributed after the session, but without letting the audience satisfy their curiosity immediately. You must, naturally, be certain to have sufficient copies for everyone in the audience.

If you decide to reproduce an 8½" × 11" hard copy of some of your presentation visuals to be distributed as handouts, make copies *only* of those visuals that are vital for later reference. While many presenters routinely distribute handouts of all their visuals, this practice can be expensive. Furthermore, the larger the package of reproduced visuals, the more likely it is to be relegated to the bottom drawer or the circular file. Conversely, the smaller the package and the more meaningful the items in it, the greater the likelihood there is of its being studied and actively referred to by audience members.

Since an effective visual should represent only key concepts from your presentation and should require further interpretation by the presenter, it will probably seem sketchy by itself. Be careful not to compromise a visual by making it overly detailed merely to improve its value as a handout. Rather, try putting a short commentary on the back of each visual distributed as a handout. This will help the audience recall the substance of your remarks and may lessen their need to take notes.

The key to success with handouts is to be certain they will make a significant contribution to the accomplishment of your objectives. If not, don't use them.

TEXTBOOKS, WORKBOOKS, AND MANUALS

Books or pamphlets can help the audience follow a procedure step-by-step, understand the rationale behind the procedure, and recall the finer points of the procedure for future reference. Again, unless you will be using these items actively during the presentation, don't make them available until *after* you have delivered your message.

OBJECTS

Often, neither a picture nor a drawing is adequate for achieving presentation results. Some situations call for the real thing. If, for example, your presentation goal is to demonstrate the capabilities of a new generation of computer software, it might be appropriate for your audience to actually work with the software, which would necessitate your providing both the computer hardware on which to run the program and copies of the software being studied.

When it is not possible to include the real thing as a presentation aid, a model can often be substituted. A model clarifies working relationships between operating components, helps with identification of parts, and provides practice in assembly and disassembly.

The usual rule about not distracting your audience with presentation aids applies to models and other items: they should not be displayed until they are to be used and should be removed or covered immediately afterward. It is also best not to pass an object around while you continue talking. This practice only ensures that you will lose the attention of whichever audience member is currently examining the item. Either focus everyone's attention on the object at once, by showing it to the group as a whole during the presentation, or let audience members examine the object individually after the session.

Careful advance planning and practice are a must before either models or actual items are used in a presentation.

Practice! Practice! Practice!

An all-too-common fault of presenters is that they don't rehearse as much as they should. Back in chapter 2, practice was discussed as one of the six principal steps in preparing any presentation. When it comes to the use of presentation aids, the need for practice cannot be overemphasized. Even if its content is top-notch, problems with the design of your presentation aids or your handling of equipment could spell the downfall of your presentation.

Practice is the only way to work the bugs out of your use of slides, projectors, chalkboards, and other presentation aids. Once you can handle them smoothly and comfortably, so they don't interfere with the rhythm of your presentation, you will be able to focus your undivided attention on achieving presentation objectives.

You will have a number of goals in rehearsing your use of presentation aids. You will, of course, be familiarizing yourself with any necessary equipment, so you can operate it without difficulty. You will be developing your ability to stand and move around in a natural manner without blocking a presentation aid from view. (This will include standing to one side of an aid and using a pointer rather than stepping in front of or otherwise obstructing the aid to call the audience's attention to something.) You will also practice maintaining eye contact and rapport with the audience while working with whatever presentation aids you have selected. In the final analysis, you will be attempting to integrate your content with your use of presentation aids to ensure that the aids and your handling of them enhance your presentation rather than compete with it.

If you take advantage of the video technology discussed earlier in this chapter to record your practice sessions, this will give you an excellent opportunity to step back and critique your handling of presentation aids and make appropriate changes. In assessing this aspect of your videotaped practice session, look for areas where the presentation aids call attention to themselves, or where your handling of them calls attention to you, rather than to the content. These are the areas that will require special attention.

In practicing your use of presentation aids, it is often important for you to try the aids out at the presentation site itself with the same equipment you will be using during the presentation—and do so far enough in advance to permit you to make any necessary modifications. A dry run at the actual location may disclose such glitches as these:

▲ Slides are an integral component of your presentation, but you find during rehearsal that the thin curtains in the meeting room you reserved don't darken the room nearly enough.

▲ The closest electrical outlet is much too far from where you'd planned to set up the overhead projector. (You did remember to pack an extension cord in your emergency kit, didn't you?)

▲ The material you intended to display on a chalkboard turns out to be too long in relation to the size of the chalkboard. You are forced either to write much smaller (thereby risking that the audience won't be able to see your material) or to choose an alternate presentation medium (perhaps flipchart pages, which could be posted on the wall—remaining visible while you continue writing additional material). (And speaking of posting flipchart pages on the wall—have you checked out whether masking tape will in fact stick to the surface of the walls at your presentation site and/or whether you are allowed to post flipchart pages on the wall?)

▲ You discover that your meeting room features a fixed screen in an awkward location.

▲ The full-scale mock-up of your new product has been placed for safekeeping in a closet in the presentation room. When you go to fetch the mock-up for display during the practice session, you discover the closet is locked, and you haven't a clue as to who has the key.

▲ While rehearsing your presentation, you find yourself fumbling constantly with the overhead transparencies and have difficulty keeping them in order—partly because the stand you are using doesn't have enough room for you to lay them out comfortably.

Obviously, the list of potential snafus involving presentation aids could be endless, but you get the idea. And many of these difficulties only reveal themselves in the course of *hands-on experience* in using the materials and equipment.

And if you can't practice at the actual location where the presentation will be given (because, for instance, the presentation will be made off-site a considerable distance away), you *can* think through the particular presentation aids you will be working with to identify critical factors and potential problem areas. With those in mind, you can then contact a knowledgeable individual at the presentation site and obtain enough information to help you avoid the pitfalls.

In short, presentation aids that turn out in practice to have been ineffectively designed or that require frenzied last-minute improvisation or involve constant fumbling constitute time wasters and distractions. At best, such *preventable problems* will give a poor impression of you and of the effort that has gone into preparing your presentation; at worst, they will keep you from achieving your objectives.

PRACTICE EXERCISES

1. Analyze the types of aids that could support the briefing you are preparing, using the information presented in this chapter. Select at least two different aids and discuss your choice with colleagues, considering any suggestions they may have.
2. If you plan to include visuals in your presentation, make a rough sketch of how two of the finished aids should look. Show them to your fellow workers to see how effectively they communicate the main points you had intended to illustrate.
3. Identify several common themes that may occur in presentations you are likely to make. Keep your eyes open for any photos, line art, or other graphic ideas that might be appropriate for emphasizing those themes. Begin a collection for your files. Make it a practice to scan the sources to which you have access and train yourself to identify good visuals.

SUMMARY

The purpose of your presentation is to communicate your message in such a way as to achieve *results*. Supporting oral remarks with visual stimuli in the form of presentation aids has been shown to markedly increase an audience's retention of the presenter's message—*if* the presentation aids are effectively designed and are skillfully used.

Designing effective presentation aids is basically a matter of breaking your message down into main points (which shouldn't be hard once you've prepared your Preliminary Plan) and deciding how to illustrate them convincingly. The technique of storyboarding is used to systematically analyze the content of an entire presentation and convert it into a series of accompanying aids.

Your best source of "raw material" for designing presentation aids could turn out to be the artwork file you yourself maintain. By saving appropriate clippings, constantly being on the lookout for artwork sources, and jotting down ideas for potential visuals, you can develop a collection of resource material tailored to precisely the type of presentations you will most often be called on to make.

In designing and selecting presentation aids, your emphasis should be on ensuring that they are clear and appropriate, have an arresting quality, are not too numerous for the length of your presentation, and really serve an auxiliary function (supporting rather than overshadowing your content).

However you select to display your presentation aids, you must be certain that the material is clearly visible or audible, that you can utilize it smoothly, and that the means to use the aid can be made readily available or accessible at the presentation site.

While many presenters rely on overhead projection, slides, and video, less technologically sophisticated media (such as audiocassette players, chalkboards, flipcharts, and handouts) still play a vital support role. The key to choosing among the many available options is determining which medium will best convey your message to your audience at your presentation site—that is, the one that will work most effectively in *your* unique combination of circumstances.

Regardless of how simple your presentation aids may seem or how familiar you think you are with them, there is no substitute for adequate rehearsal. Nor is there any substitute for well-thought-out preventive measures, backed up by a well-stocked emergency kit.

The fundamental principle governing the use of presentation aids is never to lose sight of the fact that, as their name suggests, they play a strictly *auxiliary role*. They are aids to, not replacements for, a well-prepared and well-delivered presentation.

Guidelines for the Effective Use of Presentation Aids

Never use a presentation aid before an audience until you have rehearsed with it.

▲ Be sure it works properly and that you know how to set it up correctly.

▲ Practice handling the aid and critique a videotape of your practice presentation or have an objective observer provide feedback on your use of the presentation aid.

Make certain the aid is a help rather than a hindrance to communication.

▲ Design aids that are simple and clear and that represent facts accurately. Then prepare the aids carefully.

▲ Demonstrate only one key concept per visual and be sure the visual conveys the idea better than speech alone could.

▲ Keep the use of text to a minimum, emphasizing pictures and graphics instead.

▲ Without overdoing it, use contrast, color, or other techniques to emphasize or clarify the main points.

Don't waste the audience's time with your presentation aids.

▲ Be sure all necessary equipment, components (slides, transparencies, handouts), and supporting materials (chalk, markers, a pointer) are available at the start of the presentation.

▲ Arrange the components of your presentation aid in the proper sequence prior to the session.

▲ Set up the equipment and adjust it as necessary (including focus, sound level, alignment of projection equipment with the screen, and so on) before the audience arrives.

▲ Wait until after the session to prepare presentation aids for removal and storage (including such tasks as rewinding cassettes, taking down flipchart pages posted on the wall, and placing objects or excess handouts in cartons).

Speak with more volume than is normally required. Project!

▲ Remember that the listeners' attention is divided between you and the presentation aid.

▲ Remember, too, that in a darkened room, more volume is required to hold attention.

Don't stand between your listeners and the visual aid.

▲ Stand to one side of the aid.
▲ Use a pointer if you must call the audience's attention to elements of the presentation aid.

Don't let visual aids distract *you*.

▲ Face and talk to the audience, not the presentation aid.
▲ Don't interrupt your talk when you change slides or handle aids.
▲ Use the presentation aid to support your message. Avoid modifying your message to support the aid.

Don't let visual aids distract your *audience*.

▲ Don't show a presentation aid until you are ready to use it.
▲ Handle the aid only when you are making a direct reference to it.
▲ When you are through using a presentation aid, turn it off, remove it, or cover it.
▲ Don't pass objects around during the presentation. Either show the objects to the group as a whole or display them after the session.

EFFECTIVE AIDS TO UNDERSTANDING

The following presentation aids can be used to make your topic easier to understand or more interesting and to promote the kind of thinking that will help you accomplish your objectives:

1. Charts

To direct thinking; clarify a specific point; summarize; show trends, relationships, and comparisons.

Information charts or tabulations should usually be prepared in advance as a transparency, slide, or flipchart to ensure that all points are covered and covered accurately.

The following are some of the more common types of charts:

▲ **Highlights**—to present straight copy or emphasize key points.

▲ **Time-sequence** (historical)—to show relationships over a period of time. May be in any time unit, from seconds to centuries. Can use pictures or graphs.

▲ **Organizational**—to indicate relationships between individuals, departments, sections, or jobs.

▲ **Cause-and-effect**—to illustrate causal relationships (for example, drawing of auto plus bottle [of alcoholic beverage] equals wrecked auto).

▲ **Flowchart**—to show the relation of parts to the finished whole or to the direction of movement. A PERT (program evaluation and review technique) chart is a flowchart.

▲ **Inventory**—to show a picture of an object, with its parts identified off to the side.

▲ **Dissection**—to present enlarged, transparent, or cutaway views of an object.

▲ **Diagrammatic, schematic** or **symbolic**—to provide a simplified portrayal of naturally complex objects by means of symbols (for example, a radio-wiring diagram).

▲ **Multibar graph**—to represent comparable items using horizontal or vertical bars.

▲ **Divided-bar graph**—to show the relation of parts to the whole using a single bar divided into parts by lines.

▲ **Line graph**—to display information using a horizontal scale (abscissa) and a vertical scale (ordinate) (for instance, showing the number of overtime hours being worked each month).

▲ **Divided circle** or **pie graph**—used in the same way as the divided-bar graph.

▲ **Pictograph**—to represent comparable quantities of a given item through the use of pictorial symbols (such as stacks of coins representing comparable costs of different phases of an operation).

2. Illustrations, Diagrams, and Maps

To clarify a point, emphasize trends, get attention, or show relationships or differences.

3. Video or Motion Pictures

To show movement, give an overall view or impression, or show an actual operation. Useful for affecting attitudes and emotions through various visual techniques and special effects.

4. Slides

To illustrate rules, principles, and sequences of events; to represent items as they really are (via photography).

5. Samples or Specimens

To show the real object.

6. Models

▲ **Small-scale**—to permit showing an entire operation without using large quantities of material, make a large operation visible, or show a project to be completed.

▲ **Large-scale**—to make an object big enough to permit handling, identify small parts, or show internal operation.

7. Exhibits

To show finished products, demonstrate the results of good and poor practices, attract attention, arouse and hold interest, and adequately illustrate one idea. (Use life, motion, color, or light to help attract attention.)

8. Worksheets

To provide the audience with hands-on experience in performing certain actions; to provide a carryover to the job.

9. Manuals, Pamphlets, Instruction Sheets, Circular Letters, Outlines, and Bulletins

To provide standard information and guidelines as well as reference and background material.

10. Cartoons, Posters, and Signs

To attract attention and arouse interest.

11. Photographs and Textbook or Magazine Illustrations

To tie the discussion to actual situations and people, illustrate the immediate relevance of a topic, or show local activities.

12. Case Studies

To tie together, for specific situations, the principles, practices, and procedures that are being explained, interpreted, or formulated by the group. It is much easier to visualize a procedure if you "Take the case of Mr./Mrs. . . . ."

13. Examples and Stories

To relieve monotony or tension, fix an idea, get attention, illustrate or emphasize a point, clarify a situation, or break away from a delicate or ticklish subject.

14. Demonstrations

To show how to carry out a suggested method or procedure.

15. Field Trips

To present a subject in its natural setting, stimulate interest, blend theory with practical applications, and provide additional material for study.

4

Handling Presentation Logistics

By now, you've expended a good deal of energy developing your presentation's content and creating presentation aids that will most effectively support your message. One final hurdle remains before you can actually deliver your presentation: proper attention must be given to presentation logistics—the nitty-gritty details regarding room setup, audience notification, lighting, rest rooms, name tags, and a host of other such preliminary arrangements.

"Wait just a minute," you may reply. "It's my job to give the presentation, not play janitor or meeting coordinator. I've got enough to worry about without that. Let somebody *else* take care of the details!" This sounds like a reasonable attitude—until you stop to consider that you have a vested interest in seeing that all of the details are handled properly.

If the room setup hampers the audience's view, if a piece of audio-visual equipment isn't delivered to the meeting site, or if individuals whose presence is critical fail to attend (either because the original notification was not sufficiently persuasive or because there was insufficient follow-up after the original notification), *you* will be the one with egg on your face. If the "somebody else" who has the responsibility for logistics fails to follow through effectively, it is *your* presentation (and its objectives) that will suffer.

Admittedly, managing logistics involves a great deal of work, but this work is vital to accomplishing your objectives. Just as you, the presenter, need to be involved in planning your presentation's content and developing its accompanying aids, so it is with the handling of preliminary arrangements.

Does this mean that you must personally attend to all the minutiae? Hardly. It does mean that you must provide personal input and guidance to ensure that the right questions are being asked and the right issues are being addressed. It means that you must closely supervise and/or follow up with all the "somebody elses" who are assigned to take care of the logistical details. It means, in short, that *you* must effectively assume overall responsibility for managing this aspect of your presentation.

The balance of this chapter will address the principal types of preliminary arrangements you must deal with, including inviting audience members to attend the presentation, setting up the room, preparing for the use of equipment and presentation aids, and handling the special logistical problems that arise in making an off-site presentation. A final section touches briefly on a number of fine points that are so nearly self-explanatory as to require little or no discussion here but that could nonetheless prove embarrassing or even disastrous if overlooked.

At the end of the chapter, you will find the Presentation Logistics Checklist, a tool that can help you handle preliminary arrangements in an organized fashion and with minimal expenditure of your time and effort. The checklist ensures that you not only "cover all the bases" but do so in a way that will provide detailed information about logistical arrangements. This gives both you and those who will be assisting you a clear understanding of your requirements as well as a clear record of the delegation of responsibilities and the identity of contact persons. This checklist covers the basic issues that should receive attention prior to most presentations. You may wish to adapt it to your needs by adding other items that must be checked beforehand in connection with a particular presentation.

AUDIENCE NOTIFICATION

Have all the right people—those most directly concerned or those who can make the necessary decisions—been invited to attend the presentation? How were they notified—by letter, memo, phone call, formal announcement, word of mouth? Did the notification give enough (and correct) information about the presentation's topic, its purpose, time and location, and so forth? How persuasive was the notification? Have you received replies from those invited? Have you followed up with those who were invited but have not yet replied to your notice as requested?

For best results, you should either prepare the notice yourself or provide detailed guidance to the person who will be handling notifications. Never lose sight of the fact that attendance by the right people is critical to accomplishing your presentation objectives.

ROOM SETUP

Which of the many possible room setup options you choose will depend on the size and shape of the meeting room, the size and nature of the audience, the type of presentation and delivery method, and the kind of participation you want from audience members.

This section shows some of the more conventional seating arrangements and discusses their applications. The recommended square footage to be allowed for each participant is based on a rectangular room without visual obstructions. Additional space should be allowed if your presentation room has an unusual shape, columns, or other peculiarities that might interfere with vision. In calculating total room capacity that will be needed, be sure to allow from 40 to 100 square feet for yourself and any presentation accessories you plan to use.

Among your general considerations, regardless of the particular setup you use, will be: making certain there are enough tables and chairs for the anticipated number of attendees; deciding whether a separate table is required for special guests; and determining whether you will need a podium, a microphone, or a platform from which to speak.

Auditorium Style (8 square feet per person)

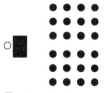

FIGURE 4-1
Auditorium Style

This arrangement, shown in figure 4-1, is useful for large groups where there is little or no need for the audience to write or consult reference materials. With this arrangement, audience participation is usually limited to question-and-answer periods.

Classroom Style (16 square feet per person)

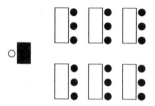

FIGURE 4-2
Classroom Style

This arrangement, shown in figure 4-2, is useful for relatively formal situations in which participants will need to write or actively use reference materials. Again, with this arrangement, audience participation is usually limited to question-and-answer periods.

Conference Style (20 square feet per person)

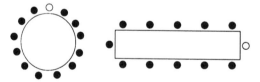

FIGURE 4-3
Conference Style (Two Options)

This arrangement, shown in figure 4-3, is useful for small groups (fifteen people or less) in which extensive discussion is desirable and in which audience members will be doing a considerable amount of writing and actively using reference materials.

Horseshoe or U-Shape Style (20 square feet per person)

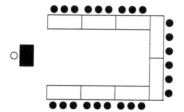

FIGURE 4-4
Horseshoe or U-Shape Style

This arrangement, shown in figure 4-4, is useful where eye contact with individual audience members, writing or use of materials by participants, and open, relatively informal discussion are all desirable.

Buzz Style (20 to 24 square feet per person)

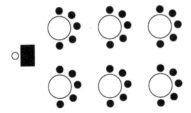

FIGURE 4-5
Buzz Style

This arrangement, shown in figure 4-5, is useful when small-group discussion will be conducted as part of the presentation. Round tables for each group are preferred.

Herringbone Style (20 square feet per person)

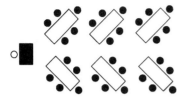

FIGURE 4-6
Herringbone Style

This arrangement, shown in figure 4-6, is also useful for group discussions. It creates a more formal air than the buzz style and a less formal air than the classroom style. Rectangular tables are used.

You could, of course, come up with many variations on these basic arrangements. The key to success in this area—whether you choose one of the standard arrangements discussed here or devise an arrangement of your own—is to utilize a setup that will best support the achievement of the objectives for your particular presentation.

ARRANGEMENTS FOR PRESENTATION AIDS AND EQUIPMENT

Are all required presentation aids and equipment in the room? Do you need a table for projection equipment or display materials? Is the equipment set up so that audience members will have an unobstructed view of the presentation aids?

Where are the electrical outlets? Will an extension cord be needed? Do the outlets accept the type of power plug on the equipment, or is an adapter needed? Is it safe to operate the equipment using an adapter or an extension cord?

Now that you've gotten the equipment plugged in, is it working properly? Do you have a spare bulb in case one blows out during the presentation? Have you set up the equipment so it is ready to present the first portion of your presentation aid? (That is, have you rewound the audio- or videocassette to the beginning, ensured that the slide

106

projector is set to display the first slide, and so on?) Have you focused the equipment?

Have you checked the room lighting to determine whether—and how—the lighting can be controlled? Do the windows have shades, blinds, or curtains to eliminate glare? And have you figured out how to operate the curtains or blinds, if the room must be darkened?

Are chalk and/or marking pens available (and in the colors you will require)? How about a chalk eraser? Have you checked to make sure the marking pens haven't dried out? Have you brought masking tape for posting flipchart pages on the walls? Have you arranged your transparencies in the proper order? Is a pointer available, if needed?

If handout materials are to be distributed, has an adequate supply been prepared? Are they assembled in the correct order and stored out of sight of the audience? Have you made provisions for distributing them at the proper time and not before?

If you plan to display objects or models, are they stored in the desired order and out of sight of the audience? If this is not possible, have you arranged for a suitable cover that can be used to shield these items from the audience's view until they are needed?

Always remember that oversights involving the preliminary arrangements for presentation aids and equipment tend to be quite readily apparent to the audience and, hence, can make you look particularly foolish. In addition to wasting time and undermining the audience's confidence in you, the spectacle you create as you hurriedly attempt to remedy this sort of problem can throw you "off your stride" for the balance of your presentation.

OFF-SITE PRESENTATIONS

Who is coordinating arrangements at the presentation site? Have you provided clear instructions as to your needs? Have arrangements been made to procure, ship, or carry with you all necessary equipment and presentation aids? If these items will arrive ahead of time, can they be stored safely until you get there? Will your travel schedule permit you to arrive early enough to make last-minute adjustments to the meeting arrangements, if necessary? If you will not be taking some of your equipment or aids with you after the presentation, have you arranged for them to be sent back to your point of origin and for them to be stored in a secure location until the return shipment can be made?

OTHER KEY ISSUES TO ADDRESS

The following list provides a brief reminder about a number of additional logistical matters that require your attention:

▲ *Reservations.* Is the planned meeting site available to you on an ongoing basis, or must it be reserved? If a reservation is required, has someone been assigned the responsibility for making it?

▲ *Seating.* We talked earlier about room setup—how the tables, chairs, and podium will be placed. Within that overall arrangement, you must be sensitive to issues of seating "protocol." Is it OK for anyone to sit anywhere, or must all seats or certain seats be reserved or assigned?

▲ *Distractions.* Have you identified potential visible or audible distractions that you must compensate for? Can arrangements be made for audience members to be given telephone messages during breaks or after the session rather than interrupting the presentation? If there is a telephone in the meeting room, have you arranged for it to be removed or shut off during the presentation? If that is not possible, have you arranged for someone not directly involved with the presentation to answer the phone and take messages?

▲ *Ventilation and smoking.* Is ventilation adequate to keep the room comfortable? Do you know where air conditioning controls are located? Do you plan to allow audience members to smoke? Will you provide a nonsmoking area? If smoking is permitted, are there enough ashtrays?

▲ *Registration.* Is there a need for participants to register? If so, how will this be handled?

▲ *Identification.* Will each audience member need to have a place card on the table in front of his or her seat? How about a stick-on or pin-on name tag? Will you put the names on these items in advance (being certain to double-check spelling), or will you provide blank cards or tags along with markers and have audience members letter their own? In fact, prior to the session, you can put out preprinted place cards on the table in front of where you want each audience member to sit as a simple means of controlling the seating arrangement.

▲ *Food, beverages, and breaks.* Has the location of rest rooms been clearly identified for participants? Will there be a short break?

Would it be desirable to provide coffee, soft drinks, or water, either during the session or during a break? Will there be a meal break? Is there a restaurant nearby? How much break time should be allotted?

The foregoing issues are neither time-consuming nor complex to deal with, and addressing them prior to the presentation can make the mechanics of your session proceed much more smoothly.

SUMMARY

While your presentations will occasionally be beset by problems over which you have no control or influence—a screen in a fixed location, a power failure, round tables where you had wanted square, a shipment of handouts lost in transit—in most cases, presentation logistics involve matters that you can directly influence. To effectively exercise control over these matters, you must keep your eyes open and your mind alert. Investing time and effort beforehand to attend to logistical details and take adequate preventive measures will pay off handsomely in the long run. And once you have the preliminary arrangements well in hand, you can concentrate on your final task—delivering the presentation.

Presentation Logistics Worksheet

Presentation topic: _____

Presenter(s): _____

Staff coordinator: _____

Date: _____ Time: _____

Address or building: _____ Room no.: _____

Site contact person: _____ Phone: _____

Attendance

No. of audience members notified: _____

No. of positive replies received: _____

Expected total attendance: _____

Room Set-up

☐ Auditorium   ☐ Conference   ☐ Buzz

☐ Classroom   ☐ Horseshoe   ☐ Herringbone

☐ Other (specify): _____

Rough-sketch the desired setup in the space provided, using the standard symbolism indicated on the left. Clearly show such relevant details as the layout of tables and chairs and the number of persons per table.

● = chair ▢ ▭ } = table ○ ■ = podium ╱ = screen ○ = presenter	

110

Room Considerations

	Status*	Responsibility	Comments
Chairs			
Tables			
Head table			
Seating protocol			
Platform			
Microphone			
Lighting			
Windows			
Distractions			
Telephone/messages			
Ventilation			
Smoking/ashtrays			
Temperature control			
Emergency exits			
Place cards or name tags			
Pencils/notepaper			
Registration			
Beverages			
Rest rooms			

*OK = satisfactory   O = not needed   + = requires attention

Presentation Aids and Equipment

	Status*	Responsibility	Comments
Needed equipment (list)			
Location of electrical outlets			
Extension cord			
Projection table			
Spare bulbs			
Supplies (chalk, marking pens, etc.) (list)			
Handouts (list)			

*OK = satisfactory   O = not needed   + = requires attention

5

Delivering the Presentation

Although this text primarily deals with planning for presentations, the actual delivery is the moment of truth. In delivering your presentation, there are a number of specific techniques you can use that will have a measurable impact on getting your message across.

In this chapter, we will start by briefly examining the nature of communication—exactly what is involved in delivering a message so that the listener will comprehend it. Then we will turn our attention to platform techniques (including eye contact, poise, gestures, facial expressions, distracting mannerisms, and so on). This is followed by a section on vocal techniques (including such speech elements as pitch, intensity, and rate as well as such voice-related pitfalls as repeated use of "uh," faulty pronunciation, and poor enunciation). Presentation tools, such as the lectern and pointer, also merit consideration, since their use can support (or detract from) presentation objectives in ways that you may not have realized. The chapter's last section deals with the techniques required to field questions from the audience in a way that enhances your overall message. (If you want to study the topic of presentation delivery in greater detail, you are encouraged to refer to some of the resources listed in the bibliography or identify additional resources through your own research.)

THE NATURE OF COMMUNICATION

Isn't it amazing how shortsighted many audiences are for not seeing your topic in the same way you do? Some of them are just like that, though, and you must learn to deal with them on that basis. And how do you do that? By *communicating* your message. (Come to think of it, if audiences weren't like that, there would probably be very little need for presentations. There might even be very little need for your services—a disturbing thought!)

To successfully communicate your message, you must realize that communication is a two-way process, as shown in figure 5-1. The presenter (P), does not communicate effectively merely by putting ideas (1) into words (2), no matter how accurate or well-chosen those words may be. The listener (L) must overcome outside interference (3) so that what is communicated not only reaches the eyes and ears (4) but enters the brain (5) in a form that is similar to the original idea (1). To test listener comprehension, the presenter must receive some form of feedback (6). Further, effective communication must lead to some form of action or response (7).

Feedback, both positive and negative, can be received in many ways: directly, through questions and comments by the listeners; and indirectly, through observation of nodding heads, facial expressions, vacant stares, attention and/or interaction or the lack thereof, and so forth. Inherent in all of these is a sensitivity on the part of the presenter—an awareness of the listeners' reactions.

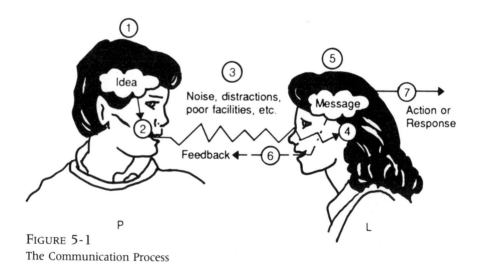

FIGURE 5-1
The Communication Process

Too often, though, in *comm_nications*, presenters leave the *you* out, forgetting that the listener is the most important component of the process. For maximum impact, the entire presentation must be listener-oriented—prepared and delivered in a manner that is understandable, interesting, and meaningful *from the viewpoint of the listener.*

PLATFORM TECHNIQUES

Platform techniques involve the presenter's nonverbal communication with the audience—how the presenter employs his or her body to support the intended message. The manner in which the presenter uses—or misuses—such techniques as eye contact, poise, gestures, and facial expressions has a significant effect on the listeners and frequently spells the difference between audience apathy and enthusiastic acceptance of the presenter's ideas. The following subsections describe several platform techniques and offer some suggestions.

Eye Contact

A vital part of effective communication is good eye contact. If you are an inexperienced presenter, maintaining eye contact with members of the audience can be both difficult and frightening. It sure is a lot more comfortable to pick out a nice, inanimate spot on the back wall and speak to it or address your presentation to the screen upon which your visual aids are being displayed! However, by doing this, you will lose out on one of the most valuable resources available for determining whether or not your message is getting across: visual feedback.

Besides, the audience will have more confidence in you if you look them in the eye when you are making your presentation. Eye contact helps you to establish a personal relationship with *each listener*. In addition, looking at a member of the audience (without staring) can be a very effective way of holding that person's attention or regaining it if it has wandered.

One very effective approach to maintaining uniform and balanced eye contact is to pick out several friendly faces in your audience, making certain that you select one from each section, and then, if necessary, address your presentation to those individuals, shifting your eyes at appropriate intervals. The effect is that you are speaking directly

115

to them (and hence to their section of the audience) during part of the presentation.

Poise and Appearance

The poised presenter is the one who looks self-confident, relaxed, and capable of doing whatever the situation may call for. Through experience, you can learn to give this impression, even when feeling insecure, by paying attention to a few rather important details:

1. Dress should be in good taste, clean, and comfortable and should be appropriate for the occasion. If informal dress is called for, use it. In more formal situations, avoid removing your suit jacket, loosening your tie, or rolling up your sleeves unless it is a working session and you want to set a mood. Use the information gathered on your Audience Analysis Audit worksheet to help you select proper clothing and accessories, keeping in mind the kind of image you want this particular audience to have of you and what you represent.
2. Approach the speaker's area in the room in a deliberate and unhurried manner. Pause for a few moments after getting into position, smile comfortably, and look your audience over briefly before beginning to speak. This gives your audience a chance to focus on you and conveys the impression that they don't frighten you (even if they do!). Also, a moment of silence will normally do far more to attract the audience's attention than launching into your presentation immediately.
3. Whether you are standing, sitting, or walking, your posture should be relaxed without being sloppy, and dignified without being stiff, reflecting the type of impression you want to give.
4. "What do I do with my hands?" is a frequent question of the less-experienced presenter. Very simply, you should do what feels most comfortable or seems most natural for you. Hold your hands loosely at your sides; raise one to your waist; put one hand in your pocket, or even both hands if you can do it without looking too casual. (Be careful not to jingle coins or keys in your pockets, though.) You may wish to fold your hands loosely in front of you, put them behind your back, rest them on the lectern or podium, or hold your notes in them.

116

The goal is to use your hands in a relaxed manner—one that does not draw attention to them at the expense of your message. And above all, avoid keeping your hands aimlessly in motion.

Gestures

Another use of your hands involves gestures. Well-selected and well-timed hand gestures can effectively promote the kind of audience reaction you want. Here are some of the more common types of gestures and the meaning they convey:

▲ *Sweeping hand* illustrates covering a broad field, takes in entire audience, and so on.
▲ *Vertical or chopping motion* emphasizes precise points and breaks an idea into parts. (Avoid waggling your index finger, though, unless you intend to scold your audience.)
▲ *Palms out* says, "Stop!" or rejects an idea.
▲ *Palms up* invites acceptance, open-mindedness, or participation.
▲ *Upturned fist* can draw the audience to you and give aggressive emphasis.

In using gestures, keep the following guidelines in mind:

▲ They should draw attention to the idea, not to the gesture itself.
▲ The types of gestures should be varied. Overuse one and it loses its effect.
▲ Properly synchronizing a gesture with the word or phrase it supports is vital.
▲ Gestures must be appropriate to the impression you want to create. The wrong gesture is worse than none at all.
▲ Using too many gestures limits their value. Control yourself if you have this tendency.

Experiment with the use of gestures and rehearse them *in private* until they become comfortable. Although they may feel awkward at first, they will get more natural with practice. Do not try out new gestures in an actual presentation (dry runs are OK) until you are reasonably confident about using them.

Body Movements

Periodic full-body movement during a presentation serves a variety of purposes. While pacing back and forth is certainly no asset, deliberate, well-timed body movements can:

▲ Relieve tension within you.
▲ Draw attention away from a visual aid and back to you.
▲ Break the hypnotic effect a stationary body has on the audience.
▲ Change the mood or pace of the presentation.

Facial Expressions

The face, in particular, should reflect the mood you want to create in your audience. The deadpan presenter will inspire neither interest nor enthusiasm. In almost any presentation, the following facial expressions will be appropriate at one time or another: serious, smiling, laughing, inquiring, doubtful. Facial expressions should be lively, varied, and appropriate to the total situation.

Distracting Mannerisms

Many people have distracting mannerisms of which they are totally unaware. Lip licking, nose patting, ear tugging, scratching, eyebrow fluttering, and head bobbing are only a few. These mannerisms can cause an audience to be fascinated, amused, or repulsed and, in so doing, can result in objectives not being achieved. You can find out whether you have any of these mannerisms by asking a sympathetic, honest coworker or by carefully reviewing a videotaped practice session. Becoming aware of the problem is half the battle. After that, it's up to you to overcome the negative habits.

Be natural! Don't try to be something or someone you aren't. Your use of nonverbal communication must be comfortable and appropriate to you, to your audience, and to your presentation topic.

VOCAL TECHNIQUES

We have all tried to listen to a speaker who was presenting interesting material and used good platform techniques but who either irritated us or put us to sleep with an unpleasant or monotonous voice. Worse than this is a speaker you either cannot hear or cannot understand. You feel that your time is being wasted and whatever he or she is mumbling about probably isn't worth much anyway.

Very few people are endowed with the voice and oratorical ability of a Franklin D. Roosevelt, a Walter Cronkite, a Billy Graham, or a Charlton Heston. Most presenters, however, can substantially increase their effectiveness in using whatever vocal tools they have if they pay careful attention to certain basic elements of speaking and if they practice. Video- or audiotaping practice sessions can also pay rich dividends.

To improve your delivery, be aware of, and learn to control, the following elements of speech:

1. *Pitch* or *inflection* refers to the tone of voice. You should strive to maintain a conversational tone that is neither too high nor too low. The tone should be natural to you and should be varied to prevent monotony. A pitch different from your normal speaking voice usually betrays nervousness and is distracting to the audience, particularly if they are familiar with the way you speak conversationally.

2. *Voice-quality* problems—a sound that is nasal, thin, harsh, pinched, or breathy—may be difficult to overcome. You can minimize the effect of such problems, though, by working with a speech instructor.

3. *Intensity* is the force or loudness with which you project. Depending on the size of the audience and the room arrangement, you should usually speak louder than you would in normal conversation. Your volume should be loud enough for everyone to hear you but not loud enough to overpower the audience. Variations in intensity can create a dynamic effect. At times, for example, a soft voice can command more attention than a loud one. If you do lower your volume, however, speak somewhat more slowly.

4. *Rate*, or tempo, of speech is another important factor. The presenter who comes across as a machine gunner usually loses an audience almost immediately, because it is impossible for them to catch everything being said. The foot dragger loses audience attention nearly as quickly by being boring or irritating, because every phrase is protracted. Again, taping your practice session can be a very helpful way to show you how good your timing is. Variations in rate can add considerably to

the effectiveness of a presentation, provided they are consistent with the kind of mood you are trying to create.

5. The *pause* is closely related to rate and can serve to draw attention to points you consider particularly important. The pause should be used deliberately, though, so as not to give the impression that you are groping for words.

The following are a few of the more common vocal problems that can really trip up your delivery:

1. The *"uh"* has long been the nemesis of public speakers. It results most often when thought processes interfere with speech processes. We fail to turn off the voice while we are thinking of what to say next. Granted, the constant use of "uh" can be extremely distracting, and anyone who has this habit should work at controlling it. Yet in our opinion, "uh" is vastly overrated as a speech problem. Although speakers have been known to come unglued at the utterance of a single such sound, the occasional "uh" in speaking is not nearly so disastrous as some would have us believe. Again, recording your practice session can be a valuable tool in helping you recognize the problem and, if necessary, overcome it.

   A technique that has helped some speakers break the "uh" cycle is to overdo it in practice. Every time you catch yourself saying "uh," repeat it two or three additional times. Overemphasis on any fault draws it more clearly to your attention. Awareness is the most important step in learning to control this problem. Complete familiarity with your presentation topic, to prevent your being at a loss for words, will also help.

2. *Voice drop* at the end of a sentence is another common fault among speakers. Without realizing it, many people let the last few words trail off to the point where it becomes difficult or impossible to hear them. Consequently, the meaning they are trying to convey is lost or distorted. Since most offenders aren't even aware they have this problem, it may take a careful review of your taped practice session to help you identify and combat it.

3. *Faulty pronunciation* is distracting to the audience and undermines their confidence in the speaker. When the audience has to take time out to identify the presenter's words, they are not going to give their undivided attention to the presentation's content. If you are not absolutely certain of the proper way to

pronounce a word, look it up in the dictionary. Mispronounced names, too, can be serious gaffes. To be sure you pronounce names correctly, check with someone who knows.

4. *Poor enunciation* means that a speaker fails to articulate the words satisfactorily. It's like talking with your mouth full. Proper enunciation, on the other hand, results in word projection that is clear, precise, and easy to listen to. The *consonants* (particularly the final ones) and not the vowels are the real keys to effective enunciation. Here again, however, keep your audience in mind and speak much as they do. Avoid sounding affected.

To correct faults in the use of your voice, you should approach the problem systematically. First, become aware of these faults by having coworkers critique your delivery or by reviewing a recorded dry run of your presentation. Study how to correct the faults, seeking help from others, if necessary. Then, practice incorrectly as well as correctly, to get a feel for how both ineffective and effective techniques sound and the difference between them.

Only you can determine how much attention you should pay to your vocal techniques, but remember that almost everyone can improve their presentations through careful attention to these basic speech elements and common problems.

PRESENTATION TOOLS

There are a number of devices that can facilitate the delivery of a speaker's message. The two most universally used are the lectern and the pointer.

Lectern or Speaker's Stand

Depending on its type and size, the lectern can serve several purposes in addition to the obvious one of providing a surface on which to place your notes:

1. It provides out-of-sight storage space for your aids and handouts, with convenient access when they are needed.
2. It gives you a resting place for your hands, although you should avoid gripping it tensely.

121

3. It serves as a tool for establishing a particular type of relationship with the audience. This is one of its most subtle yet effective uses. Remaining behind the lectern continuously tends to create a somewhat formal relationship, which can be desirable at times. Moving to the side or in front of the lectern, in addition to providing a change of pace, tends to remove both the obvious physical barrier and an unseen psychological barrier, making for a closer, more informal relationship with audience members. Moving back behind the lectern is a good way to focus attention on the summary, setting it off as a more formal part of the presentation.

Pointer

The pointer can be a valuable tool for drawing attention to specific items on a presentation aid. However, it all too frequently becomes a distracting toy. Beware of using the pointer as a pendulum or fencing foil, tapping it, or otherwise allowing it to call inappropriate attention to itself. Use the pointer only for its intended purpose, then *put it down when you're through with it!*

AUDIENCE QUESTION TECHNIQUES

Most business and governmental presentations provide for a question-and-answer period. It gives the speaker an excellent opportunity to respond to any uncertainties the audience may have and to involve the audience actively in a way that can help them mentally review and clarify the information that has been presented. Yet many an otherwise well-delivered presentation has left a poor impression precisely because of the speaker's inept handling of audience questions.

Since how you conduct the question-and-answer period can have a greater effect on the accomplishment of your objectives than the balance of the presentation, you must carefully plan exactly how and when questions will be dealt with. Furthermore, you should anticipate the types of questions and the types of questioners you may encounter and plan how you will respond to them.

In chapter 2, we discussed the audience retention curve—the typical variations in the level of an audience's attention (and hence, retention) that can be expected during the course of a presentation. Let's briefly review the topic of audience retention—this time from the standpoint of how it affects your choice of timing for a question-and-answer session.

The basic pattern of audience retention is one of heightened attention at the outset of a presentation, followed by a lower level of attention thoughout the body, and culminating with a rise in attention at the end. What this pattern tells us is that:

1. The introduction and the conclusion are the portions of a presentation that the audience is most likely to retain, so the primary points you want them to remember must be highlighted effectively here. This means that you must take as much care (if not more) in preparing the introduction and, in particular, the conclusion as you do in preparing the body of your presentation.

2. The body is the portion of a presentation that the audience is least likely to retain, despite the fact that it normally involves the most preparation effort. This does not mean that you should fail to give the body sufficient attention. Rather, it means that the body is not the place to introduce a key point and then drop it, because your audience may not remember it afterward. It also means that you should pay special attention to devising techniques that can be used to stimulate the audience during the body of a presentation in order to keep both attention and retention high. Changes of pace, audience interaction, and a wise selection of presentation aids can help.

In short, understanding the nature of communication and planning for a strong introduction, an interesting and varied development of the main ideas, and an upbeat conclusion in which you draw attention once again to the points you want the audience to remember will give your presentation every advantage. Now let's apply this information to the issue of scheduling your question-and-answer period. At what point during the presentation will a question-and-answer period have the most favorable impact?

Frequently, questions *follow* the presentation, and from the standpoint of accomplishing presentation objectives, this is often the *poorest* time for them. By turning this "prime time" over to the audience, you run the risk of having them leave the presentation remembering your difficulty in handling an embarrassing question or a curve someone may have thrown. There's also the chance that someone will take up this time asking an interesting question that is totally irrelevant to the purpose of your presentation.

Handling questions *during* a presentation can be the most effective method. This is the time when the questions are most meaningful to the people asking them. Furthermore, audience questions provide you with excellent feedback on whether or not your message is being received correctly. And since questions during the presentation require active participation by the audience, they will raise the level of retention. But there are problems, too, with taking questions midway through the presentation. If you are on a tight schedule, lengthy questions may prevent your completing the presentation on time. Unless the questions being asked are of interest to the entire audience, some listeners may feel their time is being wasted and tend to lose interest in the balance of your presentation. A premature question may also upset the way you had planned to develop your material. It takes a great deal of skill to handle questions effectively during a presentation and still maintain both control and continuity.

There is a compromise approach that can be particularly effective in those cases where established protocol dictates that the question-and-answer period come after rather than during a presentation. Consider taking questions *before you present the summary.* Since the final minutes of a presentation are "prime time" for audience retention, you can reserve that for your closing. A simple statement such as, "I'll be happy to answer any questions you may have, but I would like to hold the final two or three minutes for a summary," can give you the control you need. You can then use the summary to recover from any irrelevant, embarrassing, or critical questions and send the audience away with *your* ideas, not someone else's.

As an interest-building technique, consider leaving some important information out of the formal part of the presentation, anticipating a question from the audience on it. If a listener does ask about it, the entire audience will get a feeling of participation. At the same time, you will improve your own image by showing how effectively you can answer the question. Also, this will give you some reserve ammunition with which to counteract any opposing points of view. If your omission is not questioned by the audience, the material can be worked into your summary.

How to Conduct a Question-and-Answer Period

Since how you handle the question-and-answer period can be one of the most significant factors in the success or failure of your presentation, this subsection offers some specific pointers to keep in mind.

Your *attitude* is definitely the most important single consideration. If you approach questions as if the audience were trying to put you on the spot or catch you in a mistake, you are bound to become defensive. If, on the other hand, you approach questions as if the audience were paying you a compliment by showing their interest in the topic, the difference in climate can be tremendous. In fact, this attitude can be completely disarming to anyone who might be trying to shoot you down.

Preparedness is another key to success. You can use the Audience Analysis Audit to review who your listeners will be and consider the types of questions they might ask. Anticipating such questions beforehand gives you an opportunity to plan answers that will reinforce your presentation objectives.

Another aspect of preparedness is learning to spot ahead of time any potentially weak areas in a presentation so that you will not be unduly embarrassed if they are challenged. A government official who was requesting an increased budget to hire eight hundred additional employees did not help the cause by being unable to satisfactorily explain why the department had not filled the two hundred vacancies already in existence. Careful, objective analysis on your part should reveal most such areas of weakness. Try to study recordings of your practice sessions in order to spot places where you might be challenged or get a coworker to play devil's advocate during a dry run. Then, either develop satisfactory answers for the difficult questions you anticipate or, at the very least, decide what response you will give if such questions are raised.

Tact is an essential ability in fielding audience questions. It often becomes necessary to clarify exactly what information the questioner desires or to check your own understanding of the question. Without embarrassing the individual, you can repeat or paraphrase the question to be sure of its true meaning. If you get a buckshot question (several questions in one), you should either zero in on one precise point yourself or ask your questioner to indicate more specifically what he or she is referring to. Narrowing the focus of the question to the point where you can provide a crisp, lucid, and detailed response is much better than attempting to answer an entire broad question with vague generalities.

Phrasing your responses in terms of their *relevance* to presentation objectives is another crucial skill. You may even have to pause for a few moments to think about a question before attempting to answer it. The ability to make a quick mental evaluation and then provide an answer based on that evaluation is one of the hallmarks of an effective presenter and is a skill that requires constant, disciplined practice.

Controlling the urge to blurt out an immediate answer rather than stopping to make such an evaluation will give you a great advantage. Often, providing a detailed answer to a particular question will do little to help you accomplish presentation objectives (even if you happen to be in a position to respond in considerable detail). If such a question is asked, you should comment very briefly only on those factors that have a direct bearing on your objectives. If the questioner is not satisfied with the abbreviated answer, offer to discuss the issue in greater depth after the presentation.

And finally, you must project an air of *responsiveness* in your handling of audience questions. Always give some sort of answer, even if it's only "I don't know." Each question must be dealt with somehow or your image will suffer in the collective mind of the audience. If a question touches on a topic that will be covered later in the presentation, say so and give a condensed answer, indicating that more detail will be forthcoming. Don't postpone the answer completely unless it will seriously impair the flow of your material. In fact, coming back to the topic later will reinforce the idea you had been planning to get across. If the question is diversionary, leading the discussion away from your objectives, answer it briefly, but then take a moment to summarize where the discussion was prior to the question in order to bring your audience back on target and ensure the continuity of the presentation.

Difficult Types of Questions and Questioners

Whenever you conduct a question-and-answer period, you run a risk. You must develop skills for dealing with both difficult types of questions and difficult questioners. You should remember, however, that you have two primary responsibilities: first, to do justice to the material you are presenting and, second, to meet the needs of the *entire* audience, not just a single member of it (unless, of course, that single member is a *key individual* who must make the decisions). Here are a few suggestions for dealing with some of the problems that most often arise.

THE ARGUMENTATIVE INDIVIDUAL

The combative type, who attempts to get you into a one-on-one debate over a particular topic or point of view, must be turned aside.

In doing so, however, keep this in mind: even if you can nail that individual to the wall, you nearly always lose such an argument. First, the attacker won't let go even when nailed. Second, an extended argument is usually of little interest to the rest of the audience, and you can lose *them* in the process. Third, if you make the aggressive individual look foolish, the rest of the audience may identify with him or her and resent you for it.

Although there are exceptions, a person who argues in public is primarily seeking recognition, both from you and from the rest of the audience. An argument is an opportunity to demonstrate personal knowledge and capability or to air a particular gripe.

What's the best way to deal with an arguer? *If recognition is what is really being sought, give it and get on with your question-and-answer period.* "You've raised some very interesting ideas, Bob. I'd like to take the time to explore them in more detail with you. Can we get together right after the meeting?" You might lose a few points, but the outcome won't be nearly as disastrous as it would be if you were to stop everything and trade verbal punches.

If you do have a satisfactory answer for such a question that won't antagonize your audience, try something like: "Thanks for raising that question, Bob. I appreciate your point of view. We think that _____will take care of that problem. I'll be glad to meet with you after the session and discuss it in more detail if you wish. Now, as I was saying . . ."

THE CURVE OR LOADED QUESTION

The loaded question is specifically designed to embarrass you or put you on the spot. Frequently it can't be answered, or it hints that you are trying to hide something. A typical curve might go like this: "How do we know that you can correct these serious problems, even if you do get authorization for overtime?"

Like the arguer, this type of questioner is usually seeking recognition. He or she is, in effect, inviting you to "top this." The question is asked with the questioner knowing full well what your answer will be and having a counterquestion ready to zap you with the minute you are through delivering your response. In addition to the techniques recommended for dealing with the arguer, try pulling a turnabout on this individual: "That's a very interesting question, Barbara. What do *you* think about it?" If the game really is one-upmanship, you then have the home-team advantage. But, as with the argumentative individual, prolonging such a discussion offers no benefits. The quicker you can get off the subject, the better.

THE WINDBAG OR LONG-WINDED QUESTIONER

The rambler digresses all over the place or has to tell you a life history before getting to the point. If this type is allowed to drone on indefinitely, valuable time will be lost as well as the interest and attention of the rest of your audience. If possible, you should keep the windbag from losing face when you intervene. This can sometimes be done by putting the issue off until after the meeting. Or try one of these methods:

1. If you can anticipate the question, jump in at the first opportunity and both ask *and* answer the question for the individual.
2. Pick out a word or an idea that is being expressed and show its relationship to something you or someone else has said previously.
3. As a last resort *only,* cut the questioner off in the interests of time. Let him or her down easily and give recognition if you can in order to avoid antagonizing the individual or the rest of the audience. "That's very interesting, Connie. I wish we had the time to go into it as thoroughly as we should."

THE TAKEOVER SPECIALIST

The dominating type lets you do the work of getting an audience together, then uses the question-and-answer period to make a speech of his or her own! Whether this unsolicited soliloquy is relevant or not, the loss of control such a situation represents may hamper the achievement of your *own* objectives. Luckily, an audience is quick to note these tactics. The initial interest generated by this soapbox artist soon wanes, and the audience begins to wonder what *this* has to do with anything. "What is your question, Sam?" or "May we have your question, please" are polite but assertive responses that demonstrate to the audience not only that you are in control but that you value *their* time enough not to waste it.

WHAT TO DO IF YOU DON'T HAVE A GOOD ANSWER

If you are asked a question you can't come up with an answer for, *admit it* and either refer to someone who *can* answer or offer to find out the answer later. Bluffing is always a bad way out. Not only do you fail to satisfy the questioner but you also raise doubts in the minds of the audience as to the veracity of your entire presentation. While most of us are strongly tempted to protect our own egos by muddling through, in hopes that no one will catch our weakness, it's a temptation you must resist!

WHAT TO DO IF YOU EXPECT DIFFICULT OR COMPLEX QUESTIONS

If particularly tough questions are anticipated, you can always ask the audience to write them down on cards and send them to the front of the room at the start of the question-and-answer period. (Be sure cards and pens or pencils are available if you do this.)

If you choose to deal with questions in real time as they are asked by audience members, you may find that you need a few seconds to think, either because you have to reevaluate the question or because it catches you off guard (that is, you know the answer but need some time to collect your thoughts). *Pause.* These few moments of silence will actually enhance your credibility. Or you can try one of the following:

▲ "Would you mind repeating the question [or "restating the question in different words"] so that I can be sure I understand you?"

▲ "Your question definitely gets right to the heart of the issue. What's *your* position on that?"

▲ "That's an excellent question. Let's think about it for a few moments." (Pause until you are ready to answer.)

▲ "That question has certain interesting implications. How do some of the rest of you feel about it?"

▲ If the question lends itself to such treatment, write it or the answer on a flipchart, transparency, or chalkboard. This can provide you with some thinking time.

Much more could be written on this most critical and controversial aspect of effective presentations. The important point to remember is that the audience must understand that you welcome questions as a chance to make sure they fully comprehend both the content of your presentation and the actions expected of them.

SUMMARY

While effective platform, vocal, and audience-question techniques will not make a good presentation out of poor material, they can ensure that you deliver good material in such a way as to achieve optimal results.

Effective delivery is based on an understanding of the nature of communication—the process of delivering a listener-oriented message that has the greatest probability of eliciting the communicator's desired response.

In delivering the message, the presenter must skillfully employ such platform techniques as eye contact, poise, and body movements while avoiding distracting mannerisms, and also control such speech elements as pitch, rate, and intensity while overcoming any vocal problems (voice drop, faulty pronunciation, the "uh" habit, and the like). In addition, presentation tools, such as a speaker's stand or pointer, must be used smoothly. The goal is for the presenter to employ his or her body, voice, and presentation tools in ways that enhance communication of the intended message rather than distract listeners.

One aspect of delivery that requires special attention from the presenter is the handling of questions from the audience. The audience retention curve would suggest that a question period is most effective if scheduled during the body of a presentation, but many presenters have difficulty maintaining continuity and keeping the session on schedule if this is done. The traditional approach of taking questions at the very end results in the presenter's relinquishing control over the final portion of the presentation ("prime time" in terms of audience retention). If handling questions toward the latter part of a presentation rather than during the session seems preferable, the drawbacks of an end-of-presentation question period can largely be overcome if the presenter takes questions just before delivering the summary, thus retaining full control over the last portion—and over the audience's final impression of the subject matter.

In dealing with an audience, the presenter has to display a receptive attitude toward questions, be thoroughly prepared for areas about which he or she is likely to be questioned or challenged, be tactful in dealing with questioners, assess relevance to presentation objectives before replying to questions, and be forthcoming and appropriately responsive to all questions. Beyond these skills, the presenter must give some thought to how to deal with certain common difficulties presented by questioners (including those who are argumentative, ask loaded questions, ramble on interminably, or attempt to dominate the proceedings). This includes dealing with tough questions (both those for which the presenter has a good answer, although a certain amount of reflection may be needed before he or she delivers the reply, and those for which the presenter draws a blank).

Since the handling of the question-and-answer period can have a disproportionate impact on the achievement of presentation objectives, this aspect of the briefing deserves as much of your time and effort as the introduction, body, and conclusion.

6

Conclusion

To plan a business or technical presentation—an *effective* one, that is—start with results. What are you trying to accomplish? And is a presentation the best way to accomplish it?

Broadly speaking, there are four types of presentations: *persuasive,* which is designed to bring an audience around to your point of view on a particular topic; *explanatory,* which familiarizes an audience with a new topic, usually in general terms; *instructional,* which actively teaches something to an audience, usually in detail; and an *oral report,* which updates the audience on a matter with which they are already familiar.

Whatever the specific results sought from your presentation, a systematic, organized approach to preparing and delivering your material, such as the one illustrated in chapter 1 (repeated here as figure 6-1), is the best way to proceed. Using resource materials selected with the intended audience in mind, you develop content designed to be meaningful to that audience. In so doing, you create presentation aids that will support your message, and you give appropriate attention to preliminary logistical details and to your presentation techniques, the mechanics that affect how smoothly your session will go. Throughout, effective communication is the name of the game—getting your message across in such a way that the audience takes whatever action you intend.

Preparing your presentation essentially involves managing an investment of resources (ideas, time, energy) to achieve a desired result.

131

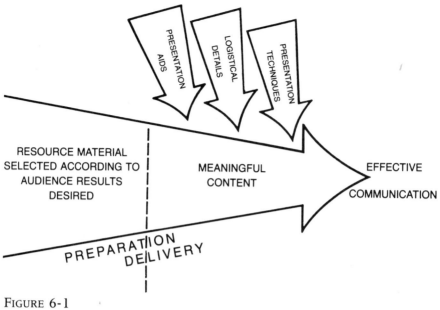

FIGURE 6-1
Elements of a Presentation

The presentation process encompasses the following six steps:

1. **Establish objectives** for the presentation. Determine *why* you are giving the presentation—that is, *what you hope to accomplish* with it. Determine whether the objectives are realistic—in terms of their scope; in terms of the audience's knowledge, background, and ability to take action; and in terms of what you can reasonably expect to achieve. Also give thought to any secondary objectives you might have (ancillary expectations, such as establishing personal credibility, that may not be openly communicated).

2. **Analyze your audience** in terms of their knowledge, attitudes, and ability to act. Tailor your approach accordingly so that your presentation will be most likely to accomplish your objectives *with this particular audience.*

3. **Prepare a Preliminary Plan** for the presentation. The Preliminary Plan is a conceptual framework, not a speaking outline. List the objectives you have established for your presentation, summarize the most pertinent information you have gathered about your audience, list the main ideas or concepts the audience *must* comprehend if your objectives are to

132

be met, and indicate the factual information needed to support and effectively communicate those ideas. Keep the preparation effort focused on results.

4. **Select resource material** for the presentation. This is done in light of the information identified in your Preliminary Plan. When you think you've identified all the resource material on which to base your presentation, submit it, item by item, to the "Why?" test to make *sure* it contributes significantly to achieving your objectives.

5. **Organize material** for effective delivery. This step involves preparing the presentation outline. An audience will tend to demonstrate heightened attention at the start of a presentation, considerably lessened attention throughout the midportion, and a sharp upswing in attention at the end. Your emphasis should therefore be on taking advantage of the two high points in audience retention by preparing an *introduction* that is brief and interesting and a *conclusion* that sums up your purpose and principal points in an arresting manner and makes an effective appeal for audience action. To maximize the level of audience involvement and attention during the *body* of the presentation (the low point in terms of retention), your aim will be to develop and illustrate your main ideas in a lively and varied fashion.

6. **Practice** the presentation in advance and **evaluate** for necessary modification. Systematically working the bugs out of your material is an absolutely essential aspect of preparing any presentation. Practice can mean delivering the presentation aloud by yourself; video- or audiotaping a practice session, to be critiqued by you and by others or staging a full-scale dry run, to be critiqued by you and by knowledgeable members of your practice audience. Focus on assessing your performance to identify areas in need of improvement.

Beyond just practicing for those presentations you are required to make, we strongly recommend that you seek out additional opportunities to practice speaking in front of others. Join Toastmasters*; take a public speaking course; volunteer for speaking assignments even if the thought paralyzes you. Experience is the only way to overcome those butterflies, or at least reduce their size. While even the most experienced presenters suffer some anxiety when their moment of truth

---

*Toastmasters International, 2200 N. Grand Avenue, Santa Ana, California 92711

is at hand, it is their experience that keeps them from letting their anxiety get in the way of their objectives.

Since presentation aids heighten your audience's retention rate by providing visual support for your oral remarks, they can play a significant role in communicating your message. To determine which points should be illustrated, focus on the main ideas identified in your Preliminary Plan (the ideas that the audience *must* retain if your presentation is to achieve its objectives). Storyboarding is a helpful way to systematically match these key ideas with visual concepts.

Visuals can be purchased, found, created, or any combination thereof. The astute presenter will constantly be alert for sources of presentation aids relevant to his or her customary topics and will maintain an art file of raw materials and ideas for presentation aids. With such a file as a ready resource, the presenter need never start from scratch when it comes to preparing presentation aids.

Never forget that presentation aids are a means of helping you communicate your message, not a substitute for that message.

Practice plays a vital role in developing all aspects of an effective business or technical presentation. When you practice, much of your attention may be devoted to the handling of presentation aids. With sufficient rehearsal, you will become able to concentrate on delivering the message rather than on the mechanics of manipulating the presentation aids. Your audience will likewise become able to turn its full attention to the substance of your presentation rather than your handling of the supporting visuals.

Another aspect of planning for a presentation is handling presentation logistics. This involves such preliminary arrangements as notifying the audience, designing a room set-up that will meet your needs, taking care of the myriad details associated with the use of presentation aids and equipment, and handling the logistics peculiar to staging an off-site presentation. You may also have to attend to such matters as reserving the meeting room, contending with distractions, providing identification for participants, and arranging for refreshments and meals.

The actual delivery of your presentation—the moment all the preparatory work has been leading up to—is based on the premise that your presentation must be listener-oriented. That is, you must aim for the approach that would be most appealing, persuasive, and interesting, not from *your* personal standpoint but from the standpoint of *the audience*. The more successful you are in doing this, the greater the likelihood that the audience will take whatever action you desire.

Delivery involves a number of different skill areas—platform techniques, vocal techniques, and handling presentation aids. Platform techniques include such aspects of nonverbal communication as eye contact, appearance, gestures and body movements, facial expressions, and eliminating distracting mannerisms. Vocal techniques address such factors in oral communication as pitch, voice quality, intensity, rate, and the pause as well as avoidance of such common speech problems as overuse of "uh," faulty pronunciation, and poor enunciation.

Handling questions from the audience is another aspect of delivery that requires considerable thought and advance planning, since the impression the presenter creates during the question period can make or break the presentation.

To successfully field the audience's questions, you must start with a positive attitude toward being questioned, be thoroughly prepared for whatever questions you can anticipate beforehand, handle questions tactfully, phrase your answers so they are directly relevant to your presentation objectives, and be responsive (to the extent possible) to *all* audience questions. You will also have to hone your skills in responding to those individuals who try to start an argument with you, toss loaded questions your way, attempt to take over the floor, or engage in protracted rambling in the course of asking a question. And finally, you will have to plan how you will respond to highly complex or difficult questions and to questions for which you have no answer.

A question-and-answer period is the audience's primary opportunity to see the presenter thinking on his or her feet and speaking extemporaneously. The audience will therefore tend to regard this period as a valid test both of the soundness of the presentation content and of the presenter's mastery over that content.

Must you do everything exactly the way it says in this book in order to make successful presentations? Not necessarily. What you must do is *communicate effectively*—that is, get your message across in a manner that will accomplish your presentation objectives.

Most presenters find that the methods described in this book constitute a highly practical approach to achieving effective communication with a minimum of time, effort, and stress. These methods, in short, provide a solid foundation in the art of making business and technical presentations. As you gain experience with presentations, you may introduce helpful modifications to the techniques advocated here, and identify certain cases where it would be advisable to bend the rules we've proposed.

To this we say: If it works, go to it! If the members of your audience do what you want them to do as a result of your presentation, you have achieved effective communication—even if you have violated every principle covered in this text!

HERE'S TO SHORTER, BETTER-ORGANIZED, AND MORE EF-FECTIVE RESULTS-ORIENTED PRESENTATIONS IN *YOUR* ORGA-NIZATION!

APPENDIX

Many of the guidelines and worksheets presented throughout the book are repeated here on perforated pages, to facilitate your working with them. You may reproduce this material, provided the permission notice shown at the foot of each page is included.

Audience Analysis Audit (AAA)

(*Fill in the blanks or circle the most descriptive terms.*)

1. **Identify the objectives** of the presentation for *this* audience. What do you want the members of the audience to do as a result of the presentation?

   _____

   _____

   _____

2. **General analysis** of the members of this audience.
   a. What is their occupational relationship to you or to the organization you represent?

   | Customers | Top management | Public |
   | Coworkers | Employees | Suppliers |

   Other (describe): _____

   b. How long have they been in this relationship?

   _____

   c. What is their vocabulary understanding level?

   | Technical | Nontechnical | Generally high |
   | Generally low | Unknown | |

   d. How willing are the members of this audience to accept the ideas you will present?

   | Eager | Receptive | Neutral |
   | Slightly resistant | Strongly resistant | Unknown |

3. **Specific analysis** of the members of this audience.
   a. What is their knowledge of the subject?

   | High | Moderate | Limited |
   | None | Unknown | |

   b. What are their opinions about the subject and about you or the organization you represent?

   | Very favorable | Positive | Neutral |
   | Slightly hostile | Openly hostile | Unknown |

Reprinted with permission from *Effective Business and Technical Presentations*, 3d ed. by George L. Morrisey and Thomas L. Sechrest, Addison-Wesley, 1987.

   c. What are their reasons for attending this presentation?

_____

_____

_____

   d. List some of the advantages and disadvantages of the presentation objectives to the members of the audience as individuals.

Advantages: _____

_____

_____

Disadvantages: _____

_____

_____

4. **Information and techniques**

   a. What types of information and techniques are most likely to gain the attention of this audience?

High-tech	Statistical comparisons	Cost-related
Anecdotes	Demonstrations	

Others (describe): _____

   b. What information or techniques are most likely to get negative reactions from members of this audience?

_____

_____

_____

5. Briefly summarize the most important information from the preceding four sections.

_____

_____

_____

_____

_____

_____

Reprinted with permission from *Effective Business and Technical Presentations*, 3d ed. by George L. Morrisey and Thomas L. Sechrest, Addison-Wesley, 1987.

Guidelines for Preparing a Preliminary Plan

The Preliminary Plan should be used as a guide:

▲ For the presenter in selecting materials, keeping ideas channeled, and determining emphasis points.
▲ For support personnel who may provide backup data, prepare presentation aids, or assist in the presentation itself.

1. **Identify specific objectives for the presentation,** keeping in mind one or more of the following criteria:
   a. They should answer the question, "Why am I giving this presentation?"
   b. They should state the results desired from the presentation—in effect, completing the sentence, "I want the following things to happen as a result of this presentation: . . ."
   c. If the body of knowledge to be presented must be identified in the objectives, use a sentence such as, "I want to tell about . . . so that . . . will take place."
   d. They should take into consideration any secondary objectives that you want to accomplish with the presentation.

2. **Identify the specific audience** for whom you are designing the presentation. State in a one- or two-sentence summary pertinent information about their knowledge, attitudes, and so forth.

3. **State the main ideas or concepts** that the audience *must* comprehend if the presentation objectives are to be met. These should:
   a. Be in conclusion form and preferably in complete sentences.
   b. Definitely lead to the accomplishment of the specific objectives.
   c. Be interesting in themselves or capable of being made so.
   d. Be few in number, usually no more than five.

4. **Identify necessary factual information** to support each of the main ideas and make them comprehensible to the audience. Avoid excessive detail.

Reprinted with permission from *Effective Business and Technical Presentations*, 3d ed. by George L. Morrisey and Thomas L. Sechrest, Addison-Wesley, 1987.

Preliminary Plan Worksheet

Topic of the presentation: _____

_____

Approximate date, time, and place for the presentation: _____

_____

Who requested that the presentation be made? _____

_____

Presentation objectives (what will be the immediate results if the presentation is successful?):

1. _____

_____

_____

2. _____

_____

_____

3. _____

_____

_____

4. _____

_____

_____

Audience analysis (who are they, and what is their general knowledge of, interest in, and attitude toward the subject?):

_____

_____

Reprinted with permission from *Effective Business and Technical Presentations,* 3d ed. by George L. Morrisey and Thomas L. Sechrest, Addison-Wesley, 1987.

Main ideas or concepts that the audience must comprehend and retain if the presentation objectives are to be met:

1. _____

_____

_____

_____

_____

2. _____

_____

_____

_____

_____

3. _____

_____

_____

_____

_____

4. _____

_____

_____

_____

_____

5. _____

_____

_____

_____

_____

Factual information necessary to support the main ideas:

Idea 1

Idea 2

Idea 3

Idea 4

Idea 5

Reprinted with permission from *Effective Business and Technical Presentations*, 3d ed. by George L. Morrisey and Thomas L. Sechrest, Addison-Wesley, 1987.

Guidelines for Organizing Material for Presentation

Introduction

The introduction has three primary purposes: (1) selling the audience on listening to the presentation, (2) introducing the subject matter, and (3) establishing your personal credibility.

Suggested approaches for the introduction:

1. *Direct statement* of the subject and why it is important to the audience.

2. *Indirect opening* dealing with some vital interest of the audience that can be linked to the subject.

3. *Vivid example* or comparison leading directly to the subject.

4. *Strong quotation* related to the subject.

5. *Important statistics* related to the subject.

6. *Story* or *anecdote* illustrating the subject.

Identify the methods you will use to state the idea and purpose of your presentation.

Body

Following the *main ideas* listed in the Preliminary Plan, provide the necessary detail for audience comprehension. Use examples, reiteration, statistics, comparisons and analogies, and expert testimony as methods to present material.

Visual illustrations are one of the most important aids to support the content. Plan carefully for their use. Also, indicate how questions and/or group discussion will be handled.

Conclusion

The conclusion is critical. It provides a summary of the main ideas, a review of the purpose of the presentation, and an appeal for audience action. It is, in effect, a minipresentation in and of itself.

Reprinted with permission from *Effective Business and Technical Presentations,* 3d ed. by George L. Morrisey and Thomas L. Sechrest, Addison-Wesley, 1987.

Presentation Worksheet

Presentation topic: _____

Presenter(s): _____

Date, time, place: _____

General Considerations

1. How will the room be arranged (seating, name cards, and so on)?

   _____

   _____

2. How many are expected to attend? How and when will they be notified of the presentation? _____

   _____

   _____

3. What presentation aids will be required? Will equipment be available at the presentation site, or must someone transport it there?

   _____

   _____

   _____

4. Will handouts be used? What arrangements have to be made for them? How and when will they be distributed? _____

   _____

   _____

   _____

5. How and when will you handle audience questions? _____

   _____

   _____

Reprinted with permission from *Effective Business and Technical Presentations*, 3d ed. by George L. Morrisey and Thomas L. Sechrest, Addison-Wesley, 1987.

Presentation Outline

Time allotted	Content*	Methods, aids, examples

Introduction: sell the audience on listening, introduce the subject, establish personal credibility; *body:* develop the main idea(s); *conclusion:* summarize content, appeal for action.

Reprinted with permission from *Effective Business and Technical Presentations,* 3d ed. by George L. Morrisey and Thomas L. Sechrest, Addison-Wesley, 1987.

Presentation Outline

Time allotted	Content*	Methods, aids, examples

Introduction: sell the audience on listening, introduce the subject, establish personal credibility; *body:* develop the main idea(s); *conclusion:* summarize content, appeal for action.

Reprinted with permission from *Effective Business and Technical Presentations*, 3d ed. by George L. Morrisey and Thomas L. Sechrest, Addison-Wesley, 1987.

Presentation Evaluation Guide

Topic: _____

Presenter: _____

Evaluator: _____

Content

INTRODUCTION

1. How well did the introduction generate interest in the presentation?

   Outstanding _____   Good _____   Fair _____   Weak _____

2. Was the purpose of the presentation made clear?

   Yes _____   Somewhat _____   No _____   Not sure _____

   Comments: _____

   _____

   _____

   _____

   _____

BODY

1. Did the main ideas come through clearly?

   Yes _____   Somewhat _____   No _____   Not sure _____

2. Were the supporting factual information and any accompanying illustrations:

   Interesting?   Yes _____   Somewhat _____   No _____

   Varied?   Yes _____   Somewhat _____   No _____

   Directly related?   Yes _____   Somewhat _____   No _____

3. Was the presentation appropriate for the intended audience?

   Yes _____   Reasonably so _____   No _____   Not sure _____

   Comments: _____

   _____

   _____

   _____

Reprinted with permission from *Effective Business and Technical Presentations*, 3d ed. by George L. Morrisey and Thomas L. Sechrest, Addison-Wesley, 1987.

CONCLUSION

1. Did the conclusion summarize the main ideas and purposes?

   Yes _____   Somewhat _____   No _____   Not sure _____

2. How effective was the conclusion in encouraging action, belief, and/or understanding?

   Outstanding _____   Good _____   Fair _____   Weak _____

Comments: _____

_____

_____

_____

_____

GENERAL

1. How would you rate the content?

   Outstanding _____   Good _____   Fair _____   Weak _____

2. Do you believe that the presentation objectives were likely to be achieved?

   Yes _____   Probably _____   No _____   Not sure _____

Comments: _____

_____

_____

_____

_____

Reprinted with permission from *Effective Business and Technical Presentations*, 3d ed. by George L. Morrisey and Thomas L. Sechrest, Addison-Wesley, 1987.

Delivery

PRESENTATION AIDS

1. Were the presentation aids suited to the topic and to the audience?

   Yes _____   Reasonably so _____   No _____

2. Were they visible to everyone and easy to follow?

   Yes _____   Reasonably so _____   No _____

3. How effective was the use of presentation aids?

   Outstanding _____   Good _____   Fair _____   Weak _____

   Comments: _____
   _____
   _____
   _____
   _____

PLATFORM TECHNIQUES

1. Poise: Was the presenter in control of the situation?

   Yes _____   Reasonably so _____   No _____

2. Were posture and movements appropriate?

   Yes _____   Reasonably so _____   No _____

3. Were gestures effective?

   Good _____   Fair _____   Overdone _____   Ineffective _____

4. Was the presenter's relationship with the audience effective (for example, eye contact)?

   Outstanding _____   Good _____   Fair _____   Weak _____

   Comments: _____
   _____
   _____
   _____
   _____

Reprinted with permission from *Effective Business and Technical Presentations*, 3d ed. by George L. Morrisey and Thomas L. Sechrest, Addison-Wesley, 1987.

VOCAL TECHNIQUES (check all that apply)

1. How were pitch and voice quality?

    Good _____   Too high _____   Too low _____

    Monotonous _____   Harsh _____   Nasal _____

2. How about rate and intensity?

    Good _____   Too fast _____   Too slow _____   Too loud _____

    Too soft _____   Monotonous _____

3. Did the presenter speak clearly and distinctly?

    Yes _____   Reasonably so _____   No _____

Comments: _____

_____

_____

_____

_____

GENERAL

1. How would you rate the overall presentation?

    Outstanding _____   Good _____   Fair _____   Weak _____

2. Make any additional comments you feel would be helpful. _____

_____

_____

_____

_____

_____

_____

_____

_____

_____

_____

_____

_____

Reprinted with permission from *Effective Business and Technical Presentations*, 3d ed. by George L. Morrisey and Thomas L. Sechrest, Addison-Wesley, 1987.

Storyboard Worksheet

Key Concepts

Visual Representations

_____

_____

_____

_____

_____

_____

_____

_____

_____

_____

_____

_____

_____

_____

_____

_____

Reprinted with permission from *Effective Business and Technical Presentations,* 3d ed. by George L. Morrisey and Thomas L. Sechrest, Addison-Wesley, 1987.

Guidelines for the Effective Use of Presentation Aids

Never use a presentation aid before an audience until you have rehearsed with it.

▲ Be sure it works properly and that you know how to set it up correctly.

▲ Practice handling the aid and critique a videotape of your practice presentation or have an objective observer provide feedback on your use of the presentation aid.

Make certain the aid is a help rather than a hindrance to communication.

▲ Design aids that are simple and clear and that represent facts accurately. Then prepare the aids carefully.

▲ Demonstrate only one key concept per visual and be sure the visual conveys the idea better than speech alone could.

▲ Keep the use of text to a minimum, emphasizing pictures and graphics instead.

▲ Without overdoing it, use contrast, color, or other techniques to emphasize or clarify the main points.

Don't waste the audience's time with your presentation aids.

▲ Be sure all necessary equipment, components (slides, transparencies, handouts), and supporting materials (chalk, markers, a pointer) are available at the start of the presentation.

▲ Arrange the components of your presentation aid in the proper sequence prior to the session.

▲ Set up the equipment and adjust it as necessary (including focus, sound level, alignment of projection equipment with the screen, and so on) before the audience arrives.

▲ Wait until after the session to prepare presentation aids for removal and storage (including such tasks as rewinding cassettes, taking down flipchart pages posted on the wall, and placing objects or excess handouts in cartons).

Speak with more volume than is normally required. Project!

▲ Remember that the listeners' attention is divided between you and the presentation aid.

Reprinted with permission from *Effective Business and Technical Presentations*, 3d ed. by George L. Morrisey and Thomas L. Sechrest, Addison-Wesley, 1987.

▲ Remember, too, that in a darkened room, more volume is required to hold attention.

Don't stand between your listeners and the visual aid.

▲ Stand to one side of the aid.
▲ Use a pointer if you must call the audience's attention to elements of the presentation aid.

Don't let visual aids distract *you*.

▲ Face and talk to the audience, not the presentation aid.
▲ Don't interrupt your talk when you change slides or handle aids.
▲ Use the presentation aid to support your message. Avoid modifying your message to support the aid.

Don't let visual aids distract your *audience*.

▲ Don't show a presentation aid until you are ready to use it.
▲ Handle the aid only when you are making a direct reference to it.
▲ When you are through using a presentation aid, turn it off, remove it, or cover it.
▲ Don't pass objects around during the presentation. Either show the objects to the group as a whole or display them after the session.

Reprinted with permission from *Effective Business and Technical Presentations,* 3d ed. by George L. Morrisey and Thomas L. Sechrest, Addison-Wesley, 1987.

Effective Aids to Understanding

The following presentation aids can be used to make your topic easier to understand or more interesting and to promote the kind of thinking that will help you accomplish your objectives:

1. Charts

To direct thinking; clarify a specific point; summarize; show trends, relationships, and comparisons.

Information charts or tabulations should usually be prepared in advance as a transparency, slide, or flipchart to ensure that all points are covered and covered accurately.

The following are some of the more common types of charts:

- ▲ **Highlights**—to present straight copy or emphasize key points.
- ▲ **Time-sequence** (historical)—to show relationships over a period of time. May be in any time unit, from seconds to centuries. Can use pictures or graphs.
- ▲ **Organizational**—to indicate relationships between individuals, departments, sections, or jobs.
- ▲ **Cause-and-effect**—to illustrate causal relationships (for example, drawing of auto plus bottle [of alcoholic beverage] equals wrecked auto).
- ▲ **Flowchart**—to show the relation of parts to the finished whole or to the direction of movement. A PERT (program evaluation and review technique) chart is a flowchart.
- ▲ **Inventory**—to show a picture of an object, with its parts identified off to the side.
- ▲ **Dissection**—to present enlarged, transparent, or cutaway views of an object.
- ▲ **Diagrammatic, schematic** or **symbolic**—to provide a simplified portrayal of naturally complex objects by means of symbols (for example, a radio-wiring diagram).
- ▲ **Multibar graph**—to represent comparable items using horizontal or vertical bars.
- ▲ **Divided-bar graph**—to show the relation of parts to the whole using a single bar divided into parts by lines.

Reprinted with permission from *Effective Business and Technical Presentations*, 3d ed. by George L. Morrisey and Thomas L. Sechrest, Addison-Wesley, 1987.

▲ **Line graph**—to display information using a horizontal scale (abscissa) and a vertical scale (ordinate) (for instance, showing the number of overtime hours being worked each month).

▲ **Divided circle** or **pie graph**—used in the same way as the divided-bar graph.

▲ **Pictograph**—to represent comparable quantities of a given item through the use of pictorial symbols (such as stacks of coins representing comparable costs of different phases of an operation).

2. Illustrations, Diagrams, and Maps

To clarify a point, emphasize trends, get attention, or show relationships or differences.

3. Video or Motion Pictures

To show movement, give an overall view or impression, or show an actual operation. Useful for affecting attitudes and emotions through various visual techniques and special effects.

4. Slides

To illustrate rules, principles, and sequences of events; to represent items as they really are (via photography).

5. Samples or Specimens

To show the real object.

6. Models

▲ **Small-scale**—to permit showing an entire operation without using large quantities of material, make a large operation visible, or show a project to be completed.

▲ **Large-scale**—to make an object big enough to permit handling, identify small parts, or show internal operation.

Reprinted with permission from *Effective Business and Technical Presentations*, 3d ed. by George L. Morrisey and Thomas L. Sechrest, Addison-Wesley, 1987.

7. Exhibits

To show finished products, demonstrate the results of good and poor practices, attract attention, arouse and hold interest, and adequately illustrate one idea. (Use life, motion, color, or light to help attract attention.)

8. Worksheets

To provide the audience with hands-on experience in performing certain actions; to provide a carryover to the job.

9. Manuals, Pamphlets, Instruction Sheets, Circular Letters, Outlines, and Bulletins

To provide standard information and guidelines as well as reference and background material.

10. Cartoons, Posters, and Signs

To attract attention and arouse interest.

11. Photographs and Textbook or Magazine Illustrations

To tie the discussion to actual situations and people, illustrate the immediate relevance of a topic, or show local activities.

12. Case Studies

To tie together, for specific situations, the principles, practices, and procedures that are being explained, interpreted, or formulated by the group. It is much easier to visualize a procedure if you "Take the case of Mr./Mrs. . . ."

Reprinted with permission from *Effective Business and Technical Presentations*, 3d ed. by George L. Morrisey and Thomas L. Sechrest, Addison-Wesley, 1987.

13. Examples and Stories

To relieve monotony or tension, fix an idea, get attention, illustrate or emphasize a point, clarify a situation, or break away from a delicate or ticklish subject.

14. Demonstrations

To show how to carry out a suggested method or procedure.

15. Field Trips

To present a subject in its natural setting, stimulate interest, blend theory with practical applications, and provide additional material for study.

Reprinted with permission from *Effective Business and Technical Presentations,* 3d ed. by George L. Morrisey and Thomas L. Sechrest, Addison-Wesley, 1987.

Presentation Logistics Worksheet

Presentation topic: _____

Presenter(s): _____

Staff coordinator: _____

Date: _____ Time: _____

Address or building: _____ Room no.: _____

Site contact person: _____ Phone: _____

Attendance

No. of audience members notified: _____

No. of positive replies received: _____

Expected total attendance: _____

Room Set-up

☐ Auditorium   ☐ Conference   ☐ Buzz

☐ Classroom   ☐ Horseshoe   ☐ Herringbone

☐ Other (specify): _____

Rough-sketch the desired setup in the space provided, using the standard symbolism indicated on the left. Clearly show such relevant details as the layout of tables and chairs and the number of persons per table.

● = chair

☐ ⎫
▭ ⎬ = table
○ ⎭

■ = podium

╱ = screen

○ = presenter

Reprinted with permission from *Effective Business and Technical Presentations*, 3d ed. by George L. Morrisey and Thomas L. Sechrest, Addison-Wesley, 1987.

Room Considerations

	Status*	Responsibility	Comments
Chairs			
Tables			
Head table			
Seating protocol			
Platform			
Microphone			
Lighting			
Windows			
Distractions			
Telephone/messages			
Ventilation			
Smoking/ashtrays			
Temperature control			
Emergency exits			
Place cards or name tags			
Pencils/notepaper			
Registration			
Beverages			
Rest rooms			

*OK = satisfactory   o = not needed   + = requires attention

Reprinted with permission from *Effective Business and Technical Presentations,* 3d ed. by George L. Morrisey and Thomas L. Sechrest, Addison-Wesley, 1987.

Presentation Aids and Equipment

	Status*	Responsibility	Comments
Needed equipment (list)			
Location of electrical outlets			
Extension cord			
Projection table			
Spare bulbs			
Supplies (chalk, marking pens, etc.) (list)			
Handouts (list)			

*OK = satisfactory   o = not needed   t = requires attention

Reprinted with permission from *Effective Business and Technical Presentations*, 3d ed. by George L. Morrisey and Thomas L. Sechrest, Addison-Wesley, 1987.

BIBLIOGRAPHY

There are thousands of books, audiotapes, videotapes, films, training courses, and other products on the market related to preparing and delivering presentations. We won't attempt to list them all here. We encourage interested readers to browse in libraries, management journals, and other professional publications for further information and guidance. However, we have described here a few resources that are of general interest and that we have found especially useful.

BOOKS

1. Broadwell, Martin M. *The Supervisor as an Instructor: A Guide for Classroom Training.* 4th ed. Reading, Mass.: Addison-Wesley Publishing Company, 1984. Since many of our readers will be making presentations that are largely instructional, Broadwell's book provides significant insight into the classroom instructional role.

   For years, Marty Broadwell has been at the forefront of instructional technology as applied to supervisors and others who are not professional trainers. As such, he is one of the most respected professionals in the field. We recommend particularly the title shown here as well as one of his video-based programs, listed below.

2. Edwards, Betty. *Drawing on the Right Side of the Brain.* Boston: Houghton-Mifflin Company, 1979. This well-known paperback can help you put visualized concepts down on paper. Even if you have never been able to draw, you may find that you have real hidden talent.

3. Frank, Milo. *How to Get Your Point Across in 30 Seconds—or Less.* New York: Simon & Schuster, 1986. This little book is a veritable gold mine of concepts and techniques for hooking your audience and keeping them tuned in to your message.

4. Hays, Robert. *Practically Speaking—In Business, Industry, and Government.* Reading, Mass.: Addison-Wesley Publishing Company, 1969. Although some portions related to the use of presentation aids are a bit dated, this remains one of the most practical, readable handbooks for those in business, industry, and government who must speak occasionally but effectively. The set of checklists for special situations is particularly useful and has a wide variety of applications.

5. Knowles, Malcolm S. *Andragogy in Action.* San Francisco: Jossey-Bass Publishers, 1984.

   ———. *The Modern Practice of Adult Education: From Pedagogy to Andragogy.* New York: Cambridge Book Company, 1980. Malcolm Knowles is often referred to as "the father of adult education." His advocacy of active participation by adults in the learning process is helpful for those who plan results-oriented programs.

6. Loney, Glenn M. *Briefing and Conference Techniques.* New York: McGraw-Hill Book Company, 1959.

   Sandford, William P., and Willard H. Yeager. *Effective Business Speech.* New York: McGraw-Hill Book Company, 1960.

   In 1962, these two resources were especially helpful in formulating concepts and techniques during the development of the original training program that eventually led to the publication of the first edition of *Effective Business and Technical Presentations* in 1968.

7. Mager, Robert F. *Measuring Instructional Results.* 2d ed. Belmont, Calif.: Pitman Learning, 1982. In this book, Mager provides a practical method for evaluating the extent to which learning objectives have been achieved.

———. *Preparing Instructional Objectives.* 2d ed. Belmont, Calif.: Pitman Learning, 1980. Here Mager specifically addresses the need for and the practice of preparing *behavior*-oriented (rather than knowledge-oriented) presentation objectives.

While the two titles listed here are the most relevant to effective presentations, any book by Bob Mager is a worthwhile learning experience as well as a pure delight to read.

8. Murray, Sheila L. *How to Organize and Manage a Seminar.* Englewood Cliffs, N.J.: Prentice-Hall, 1983. While addressed primarily to those who conduct public seminars, this book is especially useful for anyone who has to either conduct or manage extended meetings for large or small groups. It includes hundreds of tips for dealing with the nitty-gritty details.

9. Qubein, Nido R. *Communicate Like a Pro.* Englewood Cliffs, N.J.: Prentice-Hall, 1983. This book, by one of the most popular professional speakers in America, is chock-full of practical techniques related to analyzing and targeting your audience and focusing on the results you want.

OTHER MEDIA

1. *Anatomy of a Presentation.* Roundtable Films. This film has stood the test of time, and it remains one of the best on the subject. It shows what a novice presenter has to go through to get ready for a presentation and does so in a manner that most viewers can identify with. It then shows the actual presentation and provides an opportunity to evaluate it.

2. *Effective Training Techniques,* with Martin M. Broadwell—a package put out by Addison-Wesley Publishing Company that consists of three videotapes, a Leader's Guide, and Participants' Workbooks. This complete three-day training program enables participants to both observe and practice actual training techniques. It is especially helpful for those who must make many in-depth presentations.

3. *Presentation Excellence,* with Walter Cronkite. CBS/Fox Video. This video-based training program is particularly useful for understanding and practicing presentation delivery. It allows the speaker to explore and critique the presentation skills of a number of well-known figures.

4. *Speak Up with Confidence,* with Jack Valenti. National Educational Media. This book and video program suggest a number of ways to achieve presentation results, helping to make public speaking more enjoyable and productive.

Beyond the materials listed here, which are of general interest, you should search for resources that will provide the specific background you are looking for. Keep your eyes and ears open, and take advantage of every opportunity to expand your knowledge about making effective presentations.